Science Research Writing

For Non-Native Speakers of English

Science Research Writing

For Non-Native Speakers of English

Hilary Glasman-Deal
Imperial College London, UK

ICP
Imperial College Press

Published by

Imperial College Press
57 Shelton Street
Covent Garden
London WC2H 9HE

Distributed by

World Scientific Publishing Co. Pte. Ltd.
5 Toh Tuck Link, Singapore 596224
USA office: 27 Warren Street, Suite 401-402, Hackensack, NJ 07601
UK office: 57 Shelton Street, Covent Garden, London WC2H 9HE

Library of Congress Cataloging-in-Publication Data
Glasman-Deal, Hilary.
Science research writing for non-native speakers of English / by Hilary Glasman-Deal.
p. cm.
Includes bibliographical references.
ISBN 978-1-84816-309-6 (alk. paper) -- ISBN 978-1-84816-310-2 (pbk : alk. paper)
1. English language--Technical English--Handbooks, manuals, etc. 2. Technical writing--Handbooks, manuals, etc. 3. English language--Textbooks for foreign speakers. I. Title.
PE1475.G57 2009
808'.0665--dc22

2009043016

British Library Cataloguing-in-Publication Data
A catalogue record for this book is available from the British Library.

First published 2010
Reprinted in India: 2018

Printed in India

Introduction: How to Use This Book

Things should be made as simple as possible, but not any simpler.
— Albert Einstein

Who is this book for?

This book is designed to help non-native speakers of English write science research papers for publication in English. However, it can also be used as a guide for native English speakers who would like support with their science writing, and by science students who need to write a Master's dissertation or PhD thesis. It is a practical, rather than a theoretical book, and is intended as a fast do-it-yourself manual for researchers and scientists.

The book is aimed at those whose English language ability is at intermediate level or above. If you have taken an IELTS test, this is equivalent to a score of above 6.0; if you have taken a TOEFL test then this is approximately equivalent to a score above 550 (paper-based test) or 91 (iBT). However, if you have managed to read this far without using a dictionary, you will be able to use this book, even if you don't understand every word.

Why do I need it?

The goal of scientific research is publication, but good scientists are not always good writers and even native speakers of English sometimes have difficulty when they write up their research. The aim of this book is to give you the information, vocabulary and skills you need quickly and easily so that you can write confidently using the style and structure you see in the journals you read.

As a science researcher, you are able to read and understand complex, high-level material in your field. However, you may find it difficult to produce written English which is at the same level as your reading. You may feel that your English writing does not represent the content of your work effectively or accurately. The aim of this book is to enable you to use your reading ability and the material you read to develop the writing skills your work requires.

Developing the skills to write up your own research is the only way to join the international science community. If you depend on English speakers to translate your writing, their translation may not represent exactly what you intended. If you depend on proofreaders to correct your English they may not notice some errors, because a sentence which is grammatically correct is still 'wrong' if it does not mean what you intended. Also, a proofreader may not check whether your writing fits the conventional 'science research' patterns. For example, you may have forgotten to justify your choice of method or explain how your results relate to your original question, and this could mean that an editor of a science journal rejects your paper as unprofessional.

Writing and publishing a research paper is the best way to get your career off the ground. If you can turn your thesis or research project into a useful paper, your CV (Curriculum Vitae) will immediately look more professional and will be more competitive internationally. You may feel that you don't have the time to improve your English, but you already know most of what you need from the reading you have done over the years. In order to write up your research for publication you don't need to learn much more English than you already know. **Science writing is much easier than it looks.**

Most science research is written according to a fairly conventional structure: first the *title*, then the *abstract*, followed by an *introduction*, after which there is a *central section* which describes what was done and what was found and then a *discussion and/or conclusion*. At the end of the paper or research article, acknowledgements and references are added. This means that the structure of a research article will be quite similar for all writers.

Because science writing is so conventional, the amount of grammar and vocabulary you need to learn is quite small. For example, the non-technical vocabulary used in scientific writing consists of a limited set of words such as *attempt, conduct, interpret, evaluate, determine, implement, formulate, classify, correlate, enhance*, which are used as a kind of 'code'. All the vocabulary you need to get started (apart from the specialised vocabulary of your field) is in this book.

What will this book teach me?

The book will show you how to discover the conventions of structure, organisation, grammar and vocabulary in science writing in your field and will provide you with the tools to write in a similar way and at a similar level. It will teach you how to turn your research into a paper that can be submitted to a professional journal. You will also be able to use most of the information in the book and all of the language and vocabulary if you are writing a thesis in English.

I have been teaching English for Academic Purposes to science students for over 30 years. For the past 15 years I have been teaching research writing in the English Language Support Programme at Imperial College, London, where I also work closely with individual research students and staff who are writing a paper or thesis. This book is based on the most useful thing I have learned: when your language skills are not perfect, organising your information in a conventional way and using conventional language are very important. If you write according to a conventional model, the reader knows what you are trying to do because the model you are following is familiar, and language errors are therefore less significant. A researcher who begins by writing according to a simple and conventional model will soon develop higher level skills for writing independently and professionally. The opposite is also true: researchers who do not begin by writing according to a conventional model are less likely to develop these skills.

How does the book work?

The strategy in this book can be summed up as follows: carefully examine good examples of the kind of writing you would like to produce, identify and master the structure, grammar and vocabulary you see in these examples and then apply them in your own writing.

The book is divided into five units, each dealing with one section of a research article. Unit 1 deals with the Introduction, Unit 2 the Methodology, Unit 3 the Results, Unit 4 the Discussion or Conclusion and Unit 5 the

Abstract and Title. Since the aim of this book is to enable you to write in a conventional way, each unit is designed to help you discover what the conventional model of that section of a research article looks like. In each unit you will also be given support on the grammar and writing skills needed to write that section of the research article and you will be guided towards the appropriate vocabulary.

Each unit is similar. The unit on Introductions, for example, begins by looking at a sample research article Introduction similar to those in science journals, then there is a Grammar and Writing Skills section designed to respond to frequently asked questions. Because you are probably working hard on your research and don't have time to do much grammar work, there are very few grammar exercises in the Grammar and Writing Skills sections. In any case, getting the answer right in a grammar exercise doesn't automatically mean you will produce the correct grammar when you write about complex topics. Answering correctly can give you a false sense of confidence and security.

After the Grammar and Writing Skills section you will create a model or template for writing Introductions using the sample Introduction, and this is followed by a detailed Key providing model descriptors, discussion and answers to questions. The unit includes extracts from real Introductions so that you can test the model and see how it works in the 'real world'. These extracts are then used to find the vocabulary which will help you operate the model successfully. This is followed by a complete list of useful vocabulary together with examples of how the words and phrases are used.

At this stage, you will have a robust model of an Introduction, a grammar guide to deal with possible problems and a list of useful vocabulary to make the model work. Towards the end of the unit, you will be ready to test what you have learned by writing an Introduction. If you have done the tasks, you should be able to put the model, the grammar/writing skills and the vocabulary together, and a perfect Introduction will write itself almost automatically! So at the end of the unit on Introductions, you will try out what you have learned: you will write an Introduction using the model and the vocabulary list and then compare it with a sample answer in the Key.

This pattern is repeated in the rest of the units. Ideally, you should work through the book and do each task. If you read the book without completing the tasks you will have an intellectual understanding of what to do but you may find it harder to put it into practice.

Do I need any other material or books?

No, but before you begin, you should collect three or four recent research papers in your field from the journals you usually read and photocopy them. You will use these as target articles to help you adapt what you learn here to your own work, and you will refer to them while reading this book to see how the things you are learning are done in your research field. Don't use chapters from books as target articles; they are not written according to the same conventional structure as research papers and so will not help you discover how a research paper or thesis in your field is written.

Your target research articles should:

- be written by a researcher/research team based at an English-speaking institution, ideally a native speaker of English.
- be reasonably short (less than 15 A4 sides including graphs and tables).
- deal with subject matter which is as close as possible to your own topic and the kind of research you are doing.
- have clearly defined Introduction, Methodology, Results and Discussion/Conclusion sections. It will help you if these are subtitled so that you can locate them easily. Note that the subtitles may vary in different fields and even in different journals in each field; for example the Methodology can be called 'Procedure', 'Materials and Methods', 'Experimental' or some other variation.

Contents

Unit 1 ✏ How to Write an Introduction

1.1 Structure

Until now, much of your science writing has focused on writing reports in which you simply described what you did and what you found. Although this will help you write the central 'report' sections (Methodology and Results) of a research paper or thesis, it doesn't prepare you for writing an Introduction to a full-length research article; this is a new task that faces you once you move on to research writing.

In practice, you will find that you need to be certain about what you have done and what you have found in order to write the Introduction, and so the best time to write it will be after you have written, or at least drafted, the report sections. However, in this book, the structure of a research article is presented in the order in which it appears in a paper/thesis so that you can trace the connections between each part and see the sequence in which information is presented to the reader.

You may want to start your Introduction by describing the problem you are trying to solve, or the aim of your work, but as you will see when you examine published work, this is not how most research papers begin — and therefore it is not the best way for you to begin. In order to help you write the Introduction to your own research, the model you build must answer the following three questions:

- How do writers normally start the Introduction?
- What type of information should be in my Introduction, and in what order?
- How do writers normally end the Introduction?

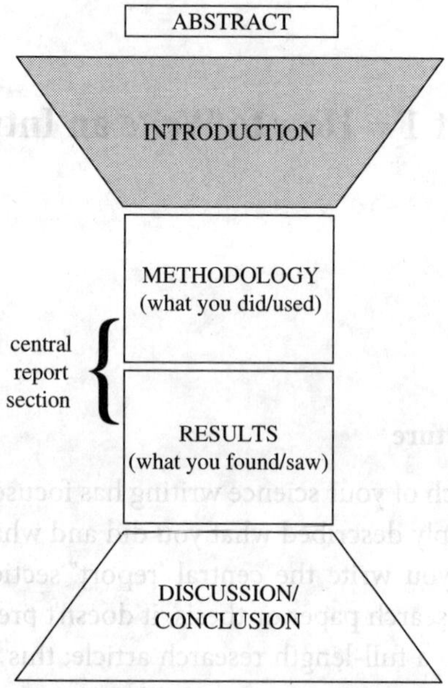

Fig. 1. The shape of a research article or thesis.

The first thing you may notice about Fig. 1 is that it is symmetrical. This is because many of the things you need to do in the Introduction are done — in reverse order — in the Discussion/Conclusion. For example, you need to write an opening sentence which enables you and your reader to 'get in' or start your paper/thesis and you also need to 'get out' at the end of the Discussion/Conclusion by finding an acceptable way to end the paper/thesis. In addition, you must look for a way to interface with the central report section at the end of the Introduction, and again — in reverse — when you move out of the central section to start the Discussion/Conclusion.

Something else you should notice about the shape of the diagram is that it narrows towards the central report section, and widens after it. This represents the way information is ordered in the Introduction and the Discussion/Conclusion: in the Introduction you start out by being fairly general and gradually narrow your focus, whereas the opposite is true in the Discussion/Conclusion.

Read the Introduction below. Don't worry if the subject matter is not familiar or if you have difficulty understanding individual words, especially technical terms like *polylactide.* Just try to get a general understanding at this stage and familiarise yourself with the type of language used.

The synthesis of flexible polymer blends from polylactide and rubber

Introduction

1 *Polylactide (PLA) has received much attention in recent years due to its biodegradable properties, which offer important economic benefits.* **2** *PLA is a polymer obtained from corn and is produced by the polymerisation of lactide.* **3** *It has many possible uses in the biomedical field*[1] *and has also been investigated as a potential engineering material.*[2,3] **4** *However, it has been found to be too weak under impact to be used commercially.*[4]

5 *One way to toughen polymers is to incorporate a layer of rubber particles*[5] *and there has been extensive research regarding the rubber modification of PLA.* **6** *For example, Penney et al. showed that PLA composites could be prepared using blending techniques*[6] *and more recently, Hillier established the toughness of such composites.*[7] **7** *However, although the effect of the rubber particles on the mechanical properties of copolymer systems was demonstrated over two years ago,*[8] *little attention has been paid to the selection of an appropriate rubber component.*

8 *The present paper presents a set of criteria for selecting such a component.* **9** *On the basis of these criteria it then describes the preparation of a set of polymer blends using PLA and a hydrocarbon rubber (PI).* **10** *This combination of two mechanistically distinct polymerisations formed a novel copolymer in which the incorporation of PI significantly increased flexibility.*

1.2 Grammar and Writing Skills

This section deals with four language areas which are important in the Introduction:

TENSE PAIRS
SIGNALLING LANGUAGE
PASSIVE/ACTIVE USE
PARAGRAPHING

1.2.1 Tense pairs

Present Simple/Present Continuous

In order to use tenses correctly in the Introduction, you first need to look at the difference between the way the Present Simple tense and the Present Continuous tense are used.

Look at these two sentences:

(a) I live in Beijing.	Present Simple
(b) I'm living in Beijing.	Present Continuous

(a) describes a permanent situation and (b) describes a temporary situation. Because of this, the Present Simple tense is used in science writing to state accepted facts and truths — but what qualifies as an accepted fact or truth is often, surprisingly, your decision. Sometimes the writer considers that research findings have the status of a fact; in that case, s/he can decide to state them in the Present Simple, usually followed by the appropriate research reference. Here is an example from the Introduction in Section 1.1:

> 5 *One way to toughen polymers* <u>*is*</u> *to incorporate a layer of rubber particles*[5] *and there has been extensive research regarding the rubber modification of PLA.*

Later on, in the Results section, you can even decide to state your own findings this way. Look at these two sentences which describe results:
(*a*) *We found that the pressure* ***increased*** *as the temperature* ***rose****, which* ***indicated*** *that temperature* ***played*** *a significant role in the process.*
(*b*) *We found that the pressure* ***increases*** *as the temperature* ***rises****, which* ***indicates*** *that temperature* ***plays*** *a significant role in the process.*

Which sentence is 'stronger'? In (a), using the Past Simple tense means that your findings are linked only to your own research, and you do not claim your deductions should be considered as accepted or established facts, or even that another researcher will necessarily get the same results. In (b), using the Present Simple tense means that you believe your findings and deductions are strong enough to be considered as facts or truths. The Present Simple communicates this reliability and your readers will respond to your work accordingly. There will be more about this later, in the unit on Results.

Past Simple/Present Perfect

Another tense pair you need in the Introduction is the Past Simple tense and the Present Perfect tense. You will need both, and you need to know when and why to switch from one to the other. Look at these sentences:

(a) Past Simple: I lived in Tokyo for five years...	but I don't live there anymore.
(b) Present Perfect: I have lived in Tokyo for five years...	and I still live there NOW.
(c) Past Simple: I broke my glasses...	but it doesn't matter/I repaired them.
(d) Present Perfect: I have broken my glasses...	and so I can't see properly NOW.

You probably learned the difference between (a) and (b) years ago: that one of the differences between Past Simple and Present Perfect is the 'time' of the verb, *i.e.* when it happened. The difference between (c) and (d) is harder to understand and more important for you as a writer of science research.

In (c) and (d), 'time', *i.e.* when the verb happened, isn't really what separates the two sentences; it's possible that both (c) and (d) happened last month, this morning, or one nanosecond ago. What is important is that the event in (d) is considered more relevant to the situation now than the event in (c), which is why it is given in the Present Perfect. Why is this idea of relevance useful when you write an Introduction? Look at these sentences from the Introduction in Section 1.1:

> *For example, Penney et al.* ***showed*** *that PLA composites could be prepared using blending techniques*[6] *and more recently, Hillier* ***established*** *the toughness of such composites.*[7] *However, although the effect of the rubber particles on the mechanical properties of copolymer systems* ***was demonstrated*** *over two years ago,*[8] *little* attention* ***has been paid*** *to the selection of an appropriate rubber component.*

* *Note*: **a little** means 'a small amount', but **little** means 'virtually none'.

Where does the tense change? Why do you think the writer changes from the Past Simple to the Present Perfect? Could it be because this research article is NOW paying attention to the selection of an appropriate rubber component?

Now look at what happens if the writer forgets to change tense and continues in the Past Simple:

> *However, although the effect of the rubber particles on the mechanical properties of copolymer systems* ***was demonstrated*** *over two years ago,*[8] *little attention* ***was paid*** *to the selection of an appropriate rubber component.*

Suddenly, the sentence means that little attention was paid THEN, *i.e.* two years ago. Perhaps attention has been paid to this problem since then; perhaps the problem has even been solved! Tense changes are always meaningful, and they always signal a change in the function of the information — so don't change tense randomly and make sure you remember to change tense when you should.

Now check what you have learned about tenses by looking carefully at the way the Past Simple and Present Perfect are used in the Introductions of your target articles. Look in particular at the way the Past Simple tense and the Present Perfect tense are used to refer to previous research.

1.2.2 Signalling language

Sentence connection

One of the most common errors in writing is failing to connect one sentence or idea to the next. Every time you end a sentence, your reader has no idea what the next sentence is going to do or say. As a result, the space between a full stop and the next capital letter is a dangerous space for you and your reader. Perhaps you stopped for ten minutes after a sentence, and during that time you thought about your work and your ideas developed. Perhaps you turned off your computer and went home. When you start typing again, if you don't share the link between those sentences with your reader, you create a gap in the text which will cause problems.

One of your tasks as a writer is to make sure that gap is closed, so that your reader is carried carefully from one piece of information to the next. Connecting sentences and concepts is good for you too, as it forces you to develop your ideas logically.

One way to connect sentences is to **overlap**, meaning to repeat something from the previous sentence:

> *The pattern of inflammation during an asthma attack is different from that seen in stable asthma. In **stable asthma** the total number of inflammatory cells does not increase.*
>
> *One way to toughen polymers is to incorporate a layer of rubber particles. As a result, there has been extensive research regarding the **rubber** modification of PLA.*

Another way is to use a **pronoun** (*it, they*) or **pro-form** (*this method, these systems*) to glue the sentences together:

> *Many researchers have suggested ways of reducing cost without affecting the quality of the image. **These methods** rely on data structures built during a preprocessing step.*
>
> *On the basis of these criteria it then describes the preparation of a set of polymer blends using PLA and a hydrocarbon rubber (PI). **This combination** of two mechanistically distinct polymerisations formed a novel copolymer in which the incorporation of PI significantly increased flexibility.*

The third way is not to finish the sentence at all, but to join it to the next sentence with a **semicolon** or a **relative clause** (a 'which' clause). Joining sentences with a semicolon works well when two sentences are very closely related and one of them is quite short:

> *The procedure for testing whether components are operationally safe usually takes many hours; this means that tests are rarely repeated.*
>
> *It has received much attention over the past few decades due to its biodegradable properties, **which** offer important economic benefits.*

The fourth way is to use a signalling sentence connector to indicate the relationship between one sentence and the next, or one part of a sentence and the next. You know how useful sentence connectors are from your reading; when you see a word like *therefore* or *however,* you are able to process the next piece of information in the sentence correctly even if you don't understand every word. This is because the sentence connector signals the function of the information in the sentence. The opposite is also true: when the writer does not signal the function of the information with a connector, it is harder for the reader to process the information. Even if the grammar is perfect and every word is correct, the reader still may not be sure what the information is doing (Is it a result of the previous sentence? An example? A cause?), and may interpret it differently from the way the writer intended.

You already use words like *therefore* and *however* and one aim of this subsection is to make sure that you are using them correctly. Another aim is to expand your vocabulary of signalling words, because you can't spend the rest of your writing life using only *therefore* and *however*! Here are some examples of signalling language arranged according to their function. It is not a long list because only those which are commonly used in science writing have been included.

CAUSE

The experiment was unsuccessful ________ the measuring instruments were inaccurate.

The experiment was unsuccessful ________ the inaccuracy of the measuring instruments.

due to (the fact that) on account of (the fact that) in view of (the fact that)	as because since

- Be careful when you use *since*; it is also often used to mean 'from that time', so if there's any possibility of confusion, choose a different connector.
- All these connectors can be used at the start of a sentence, even *because* (*Because the measuring instruments were inaccurate, the experiment was unsuccessful*).

RESULT

The measuring instruments were calibrated accurately, ________ the experiment was successful.

therefore consequently hence	as a result (of which) which is why so

- Don't start sentences with *so* to communicate a result; it's too informal.

- You can sometimes use *then,* for example in sentences like 'If x then y', but it won't work in every sentence, which is why it has not been included in this list.

CONTRAST/DIFFERENCE

British students are all vegetarians, _________ Norwegian students eat meat every day.

however whereas but	on the other hand while by contrast

- *on the contrary* and *conversely* don't fit into this category because they don't only communicate difference; they communicate the fact that 'exactly the opposite is true', so you can't use them in the sentence above (because *vegetarians* and *meat eaters* aren't opposites, they're just different). However, you could use them in the following sentence: *Some experiments used uncalibrated instruments and succeeded;* ***conversely,*** *other experiments used carefully calibrated instruments and failed.*
- Be careful when you use *while*; it is also often used to mean 'at that/the same time', so if there's any possibility of confusion, choose a different connector.

UNEXPECTEDNESS

(a) _______ it was difficult, a solution was eventually found.
(b) _______ the difficulty, a solution was eventually found.
(c) It was difficult; _______ a solution was eventually found.

(a) Although (a) Even though (a) Though	(b) Despite (b) In spite of (b) Regardless of (b) Notwithstanding	(c) nevertheless (c) however (c) yet (c) nonetheless (c) even so

- There are other connectors with the same meaning, such as *still* and *anyway,* but they are more informal.

ADDITION

We used a batch processing system because it was more effective; ___________ it was faster.

in addition	also
moreover	secondly (etc.)
furthermore	in the second place (etc.)
apart from that/which	what is more

- *besides* has more or less the same meaning as the items in the list above, but it's more powerful and is therefore better used in more persuasive contexts.

Now check what you have learned by looking at the way sentences are connected in the Introductions of your target articles.

1.2.3 Passive/Active

Students often ask whether they can use **we** in their research articles. In the Introduction you usually say what you will be doing or presenting in the research article. You can use **we** to refer to your research group or team, but do not use it to refer to people or humanity in general. If you are referring to people in general, it's better to use a construction with *It* (*It is known/ thought that...*) rather than *We know/think that...* It is also common to use the passive instead of **we**, especially in the central 'report' section (*was measured, was added, etc.*).

In a thesis, you are writing as an individual and you don't have a research group or team. Since you cannot write your thesis using **I,** you will probably write in the passive. Use words like *here* and *in this study* to let your reader know when you are referring to your own work. You can also use a 'dummy' subject to take the place of **I** or **we**:

> ***This article*** *describes an algorithm for clustering sequences into index classes.*
>
> ***The present paper*** *presents a set of criteria for selecting such a component.*

The problem with using the passive in formal writing is that the agent (the person who performed the action of the verb) is often not mentioned in the sentence. In other words, we say that something *was done* or *was identified* but we don't say 'by me' or 'by other researchers', so the reader may not know who *did* it or who *identified* it. This can cause confusion and for that reason it is sometimes clearer to use a dummy subject (*This article/ the present paper*) in the Introduction rather than the 'agentless' passive *(x is presented)*. Now look at the way the passive and dummy subject are used in the Introductions of your target articles.

PARAGRAPHING

Why is paragraphing important?
Paragraphs are an important visual aid to effective reading and writing. Two common errors in paragraphing are clusters of short or single-sentence paragraphs, and paragraphs that are too long. Both errors will confuse readers and are signs of poorly-organised writing.

To understand how paragraphing works, imagine that you have won a 24-hour trip to Paris. You have two options. The first option is to fly to Paris, get off the plane and walk around the city. If you take that option, a friend may ask you later if you saw the famous Louvre art gallery; you say: 'Well, no, I got lost and spent hours walking around the industrial area by mistake.' You show your mother the clothes you bought in Paris and she asks if you bought them in the famous Rue de la Paix shopping street, and you say, 'No, I bought them near my hotel. I didn't know where the big shopping area was.' You begin to realise that you wasted a lot of time and missed many important things.

The second option is to take a short helicopter ride over Paris before you leave the airport. It's a difficult decision because you are impatient; you only have 24 hours and you don't want to waste time, but you do it anyway. The helicopter flies over Paris for half an hour in a grid pattern, after which you begin your tour of Paris. You find a well-situated hotel, which you saw from the helicopter. You buy your clothes in the Rue de la Paix — which you saw from the helicopter. You visit the Louvre and you have lunch in one of the big parks near the centre ... which you saw from the helicopter.

What is the connection between this and good paragraphing?

Let's bring that idea to the skills of reading and writing. If you read the last page of a murder mystery before you finish the book, the rest of the story is less exciting — but you may finish the book faster. This is because you don't waste time wondering who the murderer is; you know it's the husband, so whenever his name is mentioned you concentrate and read carefully, but you don't bother to read the details about the other suspects. This enables you to read faster by giving you the confidence to ignore things which you know are not relevant.

The more you know about what you are reading, the faster and more effectively you read. So how can you find out about a long article or chapter before reading it? The answer is to skim it quickly before you begin to read. Like the helicopter ride over Paris, skimming is done before reading, not instead of reading. Your aim when you skim through a text is **to find out quickly what it is about and where the various pieces of information are located** so that you can read it faster and more confidently.

How do I skim efficiently and quickly?

Most of the instructions in the box on the next page tell you just to 'look at' or 'check' something. Skimming is a pre-reading technique and should be done very fast; if it takes more than a few minutes you're not skimming, you're reading.

Skimming may help me read, but how does it help me to write?

Look at number 6 in the box: LOOK QUICKLY AT THE FIRST SENTENCE OF EACH PARAGRAPH. A paragraph in academic writing often starts with a *topic sentence*, which gives the main idea of the paragraph, and tells the reader what the paragraph is about. The other sentences are related to

1. READ THE TITLE
   and try to predict the type of information you expect to see
2. LOOK AT THE NAME OF THE AUTHOR
   What you know about the writer will help you predict and evaluate the content.
3. CHECK THE DATE
   and use it to help you assess the content.
4. READ THE ABSTRACT
   to find out what the researchers did and/or what they found
5. LOOK QUICKLY AT THE FIRST PARAGRAPH
   without trying to understand all the words.
6. LOOK QUICKLY AT THE FIRST SENTENCE OF EACH PARAGRAPH
   without trying to understand all the words
7. LOOK QUICKLY AT EACH FIGURE/TABLE AND READ ITS TITLE
   to try and find out what type of visual data is included
8. READ THE LAST PARAGRAPH
   especially if it has a subtitle like 'Summary' or 'Conclusion'

this idea; they discuss it, describe it, define it in more detail, argue about it, give examples of it, rephrase it, *etc.* When the 'topic' or idea moves too far away from the first sentence, the writer usually begins a new paragraph.

You can therefore get a good idea of the various topics covered in an article — or in a chapter of a book — by reading the first sentence of each paragraph. And because it is a conventional way of writing paragraphs, it is a safe way for you to write paragraphs too. The more aware you are of the way other writers structure paragraphs, the easier it will be for you to do it yourself.

As you know, paragraphs are marked either by indentation (starting five spaces in) or by a double space between lines. Over the years, you have developed a very strong response to these visual signals. This means that each time you begin a new paragraph, this conditioned response in your brain prepares for a change or shift of some kind.

Correct paragraphing is essential, but it is easy to get into poor paragraphing habits, either through laziness or carelessness. If you often write one-sentence paragraphs **or** your paragraphs seem to be very long **or** you're not sure when to start a new paragraph, you are making writing harder for yourself. When you are planning your paper, write down each idea/concept that you want to talk about, checking that they are in a logical order and then listing what you want to say about each, using bullet points. This will help you create paragraphs that have a logical and coherent structure.

1.3 Writing Task: Build a Model

1.3.1 Building a model

You are now ready to begin building a model of Introductions by writing a short description of what the writer is doing in each sentence in the space provided. This may be hard, because it is the first time you are doing it, so read the guidelines below before you start. The Key is on the next page. Once you have tried to produce your own model you can use the Key to help you write this section of a research article when you eventually do it on your own.

GUIDELINES: You should spend 30–45 minutes on this task. If you can't think of a good description of the first sentence, choose an easier one, for example, Sentence 7, and start with that. Remember that your model is only useful if it can be transferred to other Introductions, so don't include content words such as *polymer* or you won't be able to use your model to generate Introductions in your own field.

One way to find out what the writer is doing in a sentence — rather than what s/he is saying — is to imagine that your computer has accidentally deleted it. What is different for you as a reader when it disappears? If you press another key on the computer and the sentence comes back, how does that affect the way you respond to the information?

Another way to figure out what the writer is doing in a sentence is to look at the grammar and vocabulary clues. What is the tense of the main verb? What is that tense normally used for? Is it the same tense as in the previous sentence? If not, why has the writer changed the tense? What words has the writer chosen to use?

Don't expect to produce a perfect model. You will modify your model when you look at the Key, and perhaps again when you compare it to the way Introductions work in your target articles.

The synthesis of flexible polymer blends from polylactide and rubber *Introduction*	**In this sentence, the writer:**
1 Polylactide (PLA) has received much attention in recent years due to its biodegradable properties, which offer important economic benefits.	1__________
2 PLA is a polymer obtained from corn and is produced by the polymerisation of lactide.	2__________
3 It has many possible uses in the biomedical field[1] and has also been investigated as a potential engineering material.[2,3]	3__________
4 However, it has been found to be too weak under impact to be used commercially.[4]	4__________
5 One way to toughen polymers is to incorporate a layer of rubber particles[5] and there has been extensive research regarding the rubber modification of PLA.	5__________
6 For example, Penney *et al.* showed that PLA composites could be prepared using blending techniques[6] and more recently, Hillier established the toughness of such composites.[7]	6__________
7 However, although the effect of the rubber particles on the mechanical properties of copolymer systems was demonstrated over two years ago,[8] little attention has been paid to the selection of an appropriate rubber component.	7__________
8 The present paper presents a set of criteria for selecting such a component.	8__________

9 On the basis of these criteria it then describes the preparation of a set of polymer blends using PLA and a hydro-carbon rubber(PI). **10** This combination of two mechanistically distinct polymerisations formed a novel copolymer in which the incorporation of PI significantly increased flexibility.	9____________ 10____________

1.3.2 Key

In Sentence 1 *'Polylactide (PLA) has received much attention in recent years due to its biodegradable properties, which offer important economic benefits.'* **the writer establishes the importance of this research topic.**

If you wrote 'introduces the topic' for Sentence 1, it won't really help when you come to write a real research article. How exactly do you 'introduce' a topic? You need to be more specific.

Most research articles begin by indicating that the research field or topic is useful or significant. They may focus on the quantity of research in this area, or how useful research in this area can be, or simply how important this research field is. If you look at your target articles, you will probably find something in the first one or two sentences that establishes the significance of the research. Phrases like *much study in recent years* or *plays a major role* are common here, and you'll find a list of useful vocabulary for this in Section 1.4.

What if I don't have the confidence to say that my research is important?

Most authors of research articles begin by establishing the significance of their research; if you don't, it can look as though your research is NOT significant, so don't be shy about stating why or how your field is important or useful.

What tense should I write in here?

Phrases like *much study in recent years* or *in the past five years* are normally followed by the Present Perfect tense (*Much study in recent years has focused on*...). Other ways of establishing significance may use the Present Simple tense (*There are substantial benefits to be gained from*...).

> **In Sentence 2** '*PLA is a polymer obtained from corn and is produced by the polymerisation of lactide.*' **the writer provides general background information for the reader.**

Sentence 2 is in the Present Simple tense, which is used for accepted/established facts (see Section 1.1). Research articles often begin with accepted or established facts. This ensures that the reader shares the same level of background information as the writer, and is therefore ready to read the article.

So what kind of facts should I start with?

This depends on how wide your subject — and therefore your readership — is. If the subject of your research is very specific, then many of your readers will have a high level of background knowledge, and you can start with fairly specific information. If your paper is likely to attract a wider audience, then you should start with more general background information. Remember that your background facts may come from research (see Section 1.1), so don't forget to include the research references where necessary.

What if there are several background facts I want to start with, not just one? How do I know which one to begin with?

Start with the most general one, the one that many of your readers will already know. This is a 'meeting place' fact, a place where all your readers can start together, after which you can move on to more specific information. Always show your readers the general picture before you examine the details: **show them the wall before you examine the bricks**! Also, don't forget to close the gap between these sentences (see Section 1.2.2) so that your readers can move smoothly through the information.

Remember that the background facts to your research are very familiar to you and the people you work with, but they won't be as familiar to all of your readers. Therefore, if the article is to reach a wider audience you need to state background facts which seem obvious or well-known to you.

I'm still not sure where to begin.

If you are still stuck for a first sentence, look at your title. It is helpful to your readers if you define the key words in your title — perhaps you can begin with a definition or a fact about one of those key words.

Can't I start by describing the problem I am hoping to solve?

You can, but most authors don't, because it's sometimes difficult to say exactly what the problem is until your readers have enough background information to understand it. It's also very hard to limit yourself to one sentence about the problem you are hoping to solve, and before you know it, you've written down a lot of specific information which your readers aren't ready for because you haven't given them enough background.

In Sentence 3 '*PLA has many possible uses in the biomedical field*[1] *and has also been investigated as a potential engineering material* [2,3]' **the writer does the same as in Sentences 1 and 2, but in a more specific/detailed way, using research references to support both the background facts and the claim for significance.**

Don't the research references mean that this is part of the literature review?

No, it's still part of the background to general research in this area. The short literature review which is generally found in the Introduction of a research article comes later, and is more likely to deal with individual studies and their methods or results. In a thesis the literature review is much longer and may be a separate chapter.

So why does the author include references if it's only the background?

For three reasons: First, because plagiarism (failing to give others the appropriate credit for their work) is unprofessional; second, referencing gives your reader the chance to find and read the study mentioned.

The third reason is that failing to provide a reference may indicate that you are not familiar with research in your area.

Although Sentence 3 isn't part of the literature review (which comes later in the Introduction) it includes a citation reference. Before you write a research paper, you collect a lot of references, quotations and ideas from journals and the Internet, many of which you will mention at some point in the paper. When you are writing the Introduction, you need to ask yourself three questions:

1. *Which of the research papers I have read should be mentioned somewhere in the Introduction?* The selection of names and references in the Introduction is important, because they draw a research 'map' for the reader by indicating the key players in your field and the progress or achievements so far. These names and references give the reader a clear idea of where your research is located and how it is related to other work in the field.

2. *Which ones should be part of the background to the research (as in Sentence 3 above) and which ones should go in the literature review which comes later in the Introduction?* If the findings are well-known and considered reliable enough to be presented as truths, you can present them in the Present Simple as part of the factual background to your paper (as in Sentence 3) with a research reference. The literature review, which describes recent and current research in your field, usually mentions authors by name, and the sentences are usually in the Simple Past or Present Perfect tense.

3. *What order should I mention them in? Who comes first and who comes last?* These questions about the literature review itself will be discussed after Sentence 6.

In Sentence 4 *'However, it has been found to be too weak under impact to be used commercially.*[4]*'* **the writer describes the general problem area or the current research focus of the field.**

Notice that the author is still not describing the specific problem which this research article will deal with; s/he is describing the current focus of the field, a problem which *many* researchers in this field are interested in and which leads to the specific problem which will be addressed in this article. Remember to keep this general description of the problem area or current research focus brief, or you will find that you begin to give a specific description of what your research is trying to achieve, and it's still too early in the Introduction for that.

As you can see from Sentence 4, you may need a research reference when you describe the problem your paper will deal with; however, if it is a well-known problem (rather than a recent issue, as in Sentence 4), then it is not necessary to provide a reference.

> **In Sentence 5** *'One way to toughen polymers is to incorporate a layer of rubber particles.*[5]*'* **the writer provides a transition between the general problem area and the literature review.**

As a general rule, you should include references to previous or current research wherever it is useful, even in a sentence whose function is primarily to provide a transition. Make sure that the superscript reference number includes all and only the work referred to in the sentence (see the notes on Sentence 6 below for more about this).

> **In Sentence 6** *'For example, Penney et al. showed that PLA composites could be prepared using blending techniques*[6] *and more recently, Hillier*[7] *established the toughness of such composites.'* **the writer provides a brief overview of key research projects in this area.**

You can't just 'pour' the literature review onto the page in any order; you should arrange your references and studies so that the reader is able to process them in a logical way. Here are three common options:

- **chronological:** Deal with the research in chronological order. This may be appropriate, for example, if the development of your field is related to political decisions.
- **different approaches/theories/models:** Group projects or studies according to their approach or methodology. Grouping similar projects together helps you avoid the 'tennis match' effect where you

go backwards and forwards, beginning each sentence in the literature review with *However* or *On the other hand*!

- **general/specific:** Start with general research in the field and gradually move to research that is closer to your own.

When should a research reference come in the middle of the sentence?

When it is necessary to avoid confusion, for example if you are referring to more than one study in a sentence or if the citation reference only refers to part of your sentence. You can see examples of this in Sentences 6 and 7.

> **In Sentence** 7 '*However, although the effect of the rubber particles on the mechanical properties of copolymer systems was demonstrated over two years ago,*[8] *little attention has been paid to the selection of an appropriate rubber component.*' **the writer describes a gap in the research**.

This is where you begin to introduce the purpose of your paper and the specific problem you will deal with, and in order to do this it is necessary to create a research space. You can do this either by describing a problem in the previous research or by indicating that there is a gap in the research. It is conventional to introduce it with a signalling connector such as *However* or *Although*. In professional writing it is unusual to put it in the form of a question; instead you can state it as a prediction or a hypothesis which you intend to test.

Don't be shy about pointing out the problems in previous research. In the first place it may be necessary in order to explain why you have done your study, and in the second place, the language used here is usually respectful and impersonal, and is therefore not considered offensive. We will look at the politeness aspect of this language in the vocabulary section at the end of the unit.

You may need more background information at this stage (for example, you may need to give details of the properties of the material which you have chosen to investigate, or describe the specific part of the device which you plan to improve). Research writing requires far more background information than you have previously given in your undergraduate writing, and it is better to offer slightly too much background information than too little.

In Sentence 8 '*The present paper presents a set of criteria for selecting such a component.*' **the writer describes the paper itself.**

At this stage you move to the present work. You can describe it, say what its purpose or focus is, give its structure or a combination of these. Check Section 1.2.3 to see whether to write these sentences in the active or the passive.

You normally use the Present Simple tense to describe the work itself (*This paper is organised as follows* or *This study focuses on*) and the Past Simple tense to talk about the aim of the work (*The aim of this project was...*), because in 'real time', the aim occurred before the work was carried out. It is also possible to state the aim in the Present Simple (*The aim of this work is...*). This is especially true in cases where the aim is only partially achieved in the paper you are submitting and the rest of the work will be done and reported on at a later stage.

In Sentence 9 '*On the basis of these criteria it then describes the preparation of a set of polymer blends using PLA and a hydrocarbon rubber(PI).*' **the writer gives details about the methodology reported in the paper.**

In Sentence 10 '*This combination of two mechanistically distinct polymerisations formed a novel copolymer in which the incorporation of PI significantly increased flexibility.*' **the writer announces the findings.**

Although you can give information about your methodology or findings in the Introduction, be careful not to go into too much detail at this point or you will find that you have nothing to write about in the Methodology or Results sections.

Look at the way the writer begins Sentences 9 and 10. In each case the information is joined to the previous sentence with a pro-form (*On the basis of* ***these criteria*** in Sentence 9 and ***This combination*** in Sentence 10).

1.3.3 The model

Here are the sentence descriptions we have collected:

In Sentence 1 the writer establishes the importance of this research topic.

In Sentence 2 the writer provides general background information.

In Sentence 3 the writer does the same as in Sentences 1 and 2, but in a more specific/detailed way.

In Sentence 4 the writer describes the general problem area or the current research focus of the field.

In Sentence 5 the writer provides a transition between the general problem area and the literature review.

In Sentence 6 the writer provides a brief overview of key research projects in this area.

In Sentence 7 the writer describes a gap in the research.

In Sentence 8 the writer describes the paper itself.

In Sentence 9 the writer gives details about the methodology reported in the paper.

In Sentence 10 the writer announces the findings.

We can streamline these so that our model has FOUR basic components:

1	ESTABLISH THE IMPORTANCE OF YOUR FIELD PROVIDE BACKGROUND FACTS/INFORMATION (possibly from research) DEFINE THE TERMINOLOGY IN THE TITLE/KEY WORDS PRESENT THE PROBLEM AREA/CURRENT RESEARCH FOCUS
2	PREVIOUS AND/OR CURRENT RESEARCH AND CONTRIBUTIONS
3	LOCATE A GAP IN THE RESEARCH DESCRIBE THE PROBLEM YOU WILL ADDRESS PRESENT A PREDICTION TO BE TESTED
4	DESCRIBE THE PRESENT PAPER

1.3.4 Testing the Model

The next step is to look at the way this model works in a real Introduction. Here are some full-length Introductions from real research articles. Read them through, and mark the model components (1, 2, 3 or 4) wherever you think you see them. For example, if you think the first sentence of the Introduction corresponds to number 1 in our model, write 1 after it, *etc.*

The height of biomolecules measured with the atomic force microscope depends on electrostatic interactions

INTRODUCTION

Because the atomic force microscope (AFM) (Binnig *et al.*, 1986) makes it possible to image surfaces in liquids, it has become an important tool for studying biological samples (Drake *et al.*, 1989). Recent reports document the observation of protein assemblies under physiological conditions at nanometer resolution (Butt *et al.*, 1990; Hoh *et al.*, 1991; Karrasch *et al.*, 1993, 1994; Yang *et al.*, 1993, Schabert and Engel, 1994; Mou *et al.*, 1995b; Muller *et al.*, 1995b, 1996b). As demonstrated on solids under vacuum conditions (Sugawara *et al.*, 1995) and in liquid (Ohnesorge and Binnig, 1993), the AFM also makes it possible to measure sample heights with subangstrom accuracy. However, the heights of native biological samples measured with the AFM in aqueous solution vary significantly, and may differ from values estimated with other methods (Butt *et al.*, 1991; Apell *et al.*, 1993; Muller *et al.*, 1995b, 1996a; Schabert and Rabe, 1996). For example, the height reported for single purple membranes ranges from 5.1 ± 0 nm to 11.0 ± 3.4 nm (see Table 1). Height measurements on actin filaments (Fritz *et al.*, 1995b), bacteriophage ø29 connectors (Muller *et al.*, 1997c), cholera toxin (Yang *et al.*, 1994; Mou *et al.*, 1995b), DNA (Hansma *et al.*, 1995; Mou *et al.*, 1995a; Wyman *et al.*, 1995), gap junctions (Hoh *et al.*, 1993), GroEL (Mou *et al.*, 1996), hexagonally packed intermediate layer (HPI) (Karrasch *et al.*, 1993; Muller *et al.*, 1996a; Schabert and Rabe, 1996), lipid bilayers (Mou *et al.*, 1994, 1995b; Radler *et al.*, 1994), and microtubules (Fritz *et al.*, 1995a) exhibit a similar variability. Height anomalies of soft surfaces have previously been studied and attributed to the mechanical properties of the sample (Weisenhorn *et al.*, 1992; Radmacher *et al.*, 1993, 1995; Hoh and Schoenenberger, 1994). However thin samples such as two-dimensional protein arrays or biological membranes adsorbed to a solid support are not sufficiently compressible to explain such large height variation.

Here we demonstrate that electrostatic interactions between the AFM tip and the sample (Butt, 1991a, b) influence the measured height of a biological structure adsorbed to a solid support in buffer solution. The DLVO (Derjaguin, Landau, Verwey, Overbeek) theory (Israelachvili, 1991) is used to describe the electrostatic repulsion and van der Waals attraction acting between tip and sample (Butt *et al.*, 1995). Experimental results and calculations show that the electrostatic double-layer forces can be eliminated by adjusting the electrolyte concentration (Butt, 1992a, b), providing conditions for correct height measurements with the AFM. In addition, the observed height dependence of the biological structure on electrolyte concentration allows its surface charge density to be estimated.

Optimal location discrimination of two multipartite pure states

1. INTRODUCTION

Entanglement lies at the heart of many aspects of quantum information theory and it is therefore desirable to understand its structure as well as possible. One attempt to improve our understanding of entanglement is the study of our ability to perform information theoretic tasks locally on non-local states, such as the local implementation of non-local quantum gates [2], telecloning [3], the remote manipulation and preparation of quantum states [4] or the recently studied question of the local discrimination of non-local states by a variety of authors. In [1] it was shown that any two orthogonal pure states can be perfectly discriminated locally, whereas in [5] examples of two *orthogonal* mixed states were presented which *cannot* be distinguished perfectly locally. Another surprising development is that there exist bases of product orthogonal pure states which cannot be locally reliably discriminated, despite the fact that each state in the basis contains no entanglement [6]. Here we discuss the issue of discriminating two non-orthogonal pure states locally, and show that in this regime the optimal global procedure can be achieved.

Inert COD production in a membrane anaerobic reactor treating brewery wastewater

INTRODUCTION

The chemical characterization of wastewaters is commonly undertaken to determine their biological treatability, load on an existing treatment system, or compliance with the final discharge standards. In each case, one of the most important parameters to be measured is the chemical oxygen demand (COD). In general, the COD value of a wastewater mainly represents the biodegradable and non-biodegradable organic components, although inorganic compounds may be significant in certain cases. In biological treatment systems, the biodegradable fraction of wastewater can be removed effectively, but its non-biodegradable fraction passes through the system unchanged. In addition to this, a significant amount of soluble microbial products may be produced by microorganisms within the treatment systems. Some of these will be resistant to biological degradation and will appear in reactor effluents. The factors that affect effluent quality and overall organic matter removal in biological treatment systems are, therefore, the presence of both the inert COD fraction in the influent wastewater and the soluble microbial products which are produced during biological treatment. Although their concentrations may have few practical implications in the treatment of low strength wastewaters, they may have relatively greater significance in the treatment of medium-high strength industrial wastewaters.

There is extensive literature on the determination of inert COD fractions in industrial wastewaters under aerobic conditions (Chudoba, 1985; Ekama *et al.*, 1986; Rittman *et al.*, 1987;

Henze *et al.*, 1987; Orhon *et al.*, 1989; Germirli *et al.*, 1991). However, little has been reported under anaerobic conditions (Germili *et al.*, 1998; Ince *et al.*, 1998). Since medium-high strength industrial wastewaters have been treated efficiently by anaerobic treatment systems, both the inert COD fraction of wastewaters under anaerobic conditions and the soluble microbial products produced within the anaerobic treatment systems should be investigated.

A novel anaerobic reactor system, crossflow ultrafiltration membrane anaerobic reactor (CUMAR) has previously shown great potential for retaining high biomass levels and high biological activity within a fully functioning anaerobic digester (Ince *et al.*, 1993, 1994, 1995a). Since the CUMAR system can be operated at high organic loading rates, the quantification of its efficiency under varying loading rates would be of considerable interest, particularly with regard to the nature and quantity of soluble COD produced in the reactor effluent under various operating conditions.

In this study, formation of soluble microbial products within a 120:1 [is this correct? Should it be 120:1?] pilot-scale CUMAR system treating brewery wastewater will, therefore, be discussed in relation to reactor operating conditions.

Organic vapour phase deposition: a new method for the growth of organic thin films with large optical non-linearities

1. INTRODUCTION

There is considerable interest in organic materials with large second-order hyperpolarizabilities for use in non-linear optical (NLO) devices such as modulators and frequency doublers [1]. To achieve a high figure of merit for such NLP devices requires a material with a non-centrosymmetric bulk structure and low dielectric constant.

To this end, NLP-active chromophores are traditionally incorporated into a polymer matrix and electrically poled to achieve the necessary bulk symmetry. However, such materials are limited by their low glass transition temperatures and poor stabilities at elevated temperature.

Recently, single crystals of organic and organometallic salts [2–4] have been shown to possess extremely large second-order ($x(^2)$) NLP effects leading to a high second harmonic generation (SHG) efficiency. The naturally non-centrosymmetric crystal structures of these compounds obviates the need for external poling. Furthermore, these salts have a high optical damage threshold and sufficient stability with respect to temperature to withstand many conventional semiconductor fabrication processes. In particular, highly pure single crystals of the salt, 4′-dimethylamino-N-methyl-4-stilbazolium tosylate (DAST) [2], have been shown to have a value of $x(^2)$ at least 10^3 times greater than that of urea due to dipole alignment of the cation and anion constituents of the DAST structure. To illustrate this alignment, the DAST bulk crystal structure is shown in the inset of Fig. 1.

For many applications such as waveguide devices, it is desirable to grow NLO materials into optical quality thin films. Although thermal evaporation in a high vacuum environment has been used to grow thin films of many organic [5–7] and inorganic materials, the technique is not always applicable to highly polar molecules [8] or organic salts.

For example, when heated in vacuum, DAST decomposes before vaporization. Although in situ reactions of multicomponent organic molecules to synthesize polymer films previously has been demonstrated using vacuum techniques as physical vapour deposition or vapour deposition polymerization [9], attempts in our own laboratory at double-source co-evaporation of DAST neutral precursors 4′-dimethylamino-4-stilbazole (DAS) and methyl p-toluenesulfonate (Methyltosylate, MT) to form DAST have been unsuccessful, due in part to the radically different vapour pressures of DAS and MT, which leads to highly non-stoichiometric growth.

In contrast, atmospheric or low pressure (eg milliTorr) vapour phase epitaxy (VPE) has been used to grow epitaxial thin films of many III-V compound semiconductors, such as InP and GaAs, where there is a large difference in the vapour pressures of the group III and group V atomic constituents [10]. This method was recently extended to allow the growth of III-V and II-VI semiconductors from volatile organic precursors [11]. Here, a high vapour pressure compound (typically a metal halide or a metallorganic) of each respective metal is carried independently, via a carrier gas, to a high temperature reaction zone. In this zone, the compounds are deposited onto a heated substrate where they thermally decompose and react to yield the desired III-V compound. The excess reactants and reaction products are then exhausted from the system via a scrubber.

In this paper we apply the techniques of VPE to grow films of DAST by the reaction of two volatile organic materials in a hot-wall, atmospheric pressure reactor. By nuclear magnetic resonance (NMR) analysis, we find that the stoichiometry of polycrystalline DAST films is >95% pure (limited by instrumental sensitivity). Using X-ray diffraction and other analytical techniques, we observe a significant dependence of film quality, such as ordering and crystallite size, on the substrate composition and other deposition conditions used for growth, suggesting that it may be possible to generate optical quality thin films of DAST and similar organic salts and compounds by OVPD using suitable substrates. To our knowledge, this is the first demonstration of the deposition of ordered thin films of a highly non-linear optically active organic salt using atmospheric vapour phase techniques.

Limitations of charge-transfer models for mixed-conducting oxygen electrodes

INTRODUCTION

Traditionally, electrochemistry is concerned with charge-transfer reactions occurring across a 2-dimensional interface. Indeed, at any macroscopic two-phase boundary, the magnitude, direction and driving force for current density can be described relatively unambiguously. As early as 1933 [1], workers began introducing the concept of a 'three-phase boundary' (solid/liquid/gas) in order to allow for direct involvement of gas-phase species at an electrochemical interface. However, since matter cannot pass through a truly one-dimensional interface among three phases, concepts of 'interfacial area', 'current density', and 'overpotential' at a three-phase boundary lack clear definition. For example, where exactly is the current flowing from/to, and what is the local flux density? Also, if we define overpotential in terms of thermodynamic potentials of species outside the interfacial region, what species and region are we talking about? Although the three-phase boundary concept may serve as a useful abstraction of the overall electrode reaction, it does not address these mechanistic questions.

Workers studying gas-diffusion electrodes in the mid-1960s recognized the limitations of the three-phase boundary concept [2, 3]. As an alternative, they began to break down the electrode reaction into individual steps, some that involve charge-transfer across a two-dimensional interface, and some that involve dissolution and diffusion of molecular species in three dimensions or across a chemical interface. These and subsequent studies have demonstrated that electrodes with *i-V* characteristics indicative of charge-transfer limitations (eg. Tafel behaviour) can, in fact, be limited by steps that do not themselves involve charge-transfer [4]. Although the solid-state literature has held on to the three-phase boundary concept more tightly than the aqueous or polymer literature, few examples remain today or solid-state electrochemical reactions that are not partially limited by solid-state reaction and diffusion processes.

One example is the O_2-reduction reaction on a mixed-conducting perovskite electrode, which defies rational explanation in terms of interfacial impedance. In order to incorporate non-charge-transfer effects, workers often apply an empirical Butler–Volmer model (for DC characteristics) or an equivalent-circuit model (for AC impedance) that treat non-charge-transfer processes in terms of an *effective* overpotential/current relationship [5, 6]. However, this approach lacks generality and can often be incorrect for treating oxygen absorption and solid-state and gaseous diffusion, which contribute to the impedance in a convoluted manner [7]. Although such models may provide a useful set of parameters to 'fit' data accurately, they leave the electrode reaction mechanism only vaguely or empirically defined, and provide little mechanistic insight.

The purpose of this paper is to provide a framework for defining 'charge-transfer' and 'non-charge-transfer' processes, and to illustrate how they are different. We investigate why charge-transfer models have difficulty modelling non-charge-transfer effects, and walk through several examples including the ALS model for oxygen reduction on a porous mixed-conducting oxygen electrode. We then review a recent study of linear AC polarization of La_{1-x} Sr_x CoO_{3-5} (LSCO) electrodes on ceria that corroborates the ALS model, and demonstrates the importance of O_2 surface exchange and diffusion. This study shows that the electrode reaction extends up to 20 microns beyond the electrode/electrolyte interface, implying that electrode polarization is better described by macroscopic thermodynamic gradients than as an 'overpotential'.

Now do the same for the Introductions of your target articles. You should find that most Introductions begin with item 1, that the order of the model components is usually fairly reliable (although items 2 and 3 can occur more than once) and that almost all Introductions finish with number 4. We have, therefore, answered the three questions we set at the beginning of this unit:

- How do I start the Introduction? What type of sentence should I begin with?
- What type of information should be in my Introduction, and in what order?
- How do I end the Introduction?

1.4 Vocabulary

You now need to collect vocabulary for each part of the Introduction model. The vocabulary in this section is taken from over 600 research articles in different fields, all of which were written by native speakers and published in science journals. Only words/phrases which appear frequently have been included; this means that the vocabulary lists contain words and phrases which are considered normal and acceptable by both writers and editors. We will look at vocabulary for the following areas of the model:

1. ESTABLISHING SIGNIFICANCE

This includes phrases such as *Much research in recent years*. A good list of commonly used words and expressions will encourage you to include this in your first sentences.

2. PREVIOUS AND/OR CURRENT RESEARCH AND CONTRIBUTIONS

This includes all past tense verbs describing what researchers did, *i.e. calculated, monitored, etc.* Instead of just using *did, showed* and *found*, you often need to be more specific about what a researcher actually 'did'!

3. GAP/PROBLEM/QUESTION/PREDICTION

This includes ways to say exactly how previous and/or current research is not yet complete or has not addressed the problem your paper deals with, *e.g. However, few studies have focused on...*

4. THE PRESENT WORK

This may include your purpose, your strategy and the design of your paper, using language such as *the aims of the present work are as follows:*

VOCABULARY TASK

Look through the Introductions in this unit and the Introductions of your target articles. Underline or highlight all the words and phrases that you think could be used in each of the four areas given above.

A full list of useful language can be found on the following pages. This includes all the appropriate words and phrases from the Introductions in this unit, together with some other common ones which you may have seen in your target articles. Underneath each list you will find examples of how they are used. Read through the list and check the meaning of any you don't know in the dictionary. This list will be useful for many years.

1.4.1 Vocabulary for the Introduction

1. ESTABLISHING SIGNIFICANCE

(a) basic issue	economically important
(a) central problem	(has) focused (on)
(a) challenging area	for a number of years
(a) classic feature	for many years
(a) common issue	frequent(ly)
(a) considerable number	generally
(a) crucial issue	(has been) extensively studied
(a) current problem	importance/important
(a) dramatic increase	many
(an) essential element	most
(a) fundamental issue	much study in recent years
(a) growth in popularity	nowadays
(an) increasing number	numerous investigations
(an) interesting field	of great concern
(a) key technique	of growing interest
(a) leading cause (of)	often
(a) major issue	one of the best-known
(a) popular method	over the past ten years
(a) powerful tool/method	play a key role (in)
(a) profitable technology	play a major part (in)
(a) range (of)	possible benefits
(a) rapid rise	potential applications
(a) remarkable variety	recent decades
(a) significant increase	recent(ly)
(a) striking feature	today
(a) useful method	traditional(ly)
(a) vital aspect	typical(ly)
(a) worthwhile study	usually

(an) advantage	well-documented
attracted much attention	well-known
benefit/beneficial	widely recognised
commercial interest	widespread
during the past two decades	worthwhile

Here are some examples of how these are used:

- A **major current focus** in population management is how to ensure sustainability of...
- **Numerous experiments have established that** ionising radiation causes...
- Low-dose responses to radiation have **generated considerable recent research interest.**
- Analysis of change in the transportation sector is **vital** for two **important** reasons: ...
- PDA accounts for **over 95%** of all pancreatic cancers.
- **It is generally accepted that** joints in steel frames operate in a semi-rigid fashion.
- Nanocrystalline oxide films **are attracting widespread interest** in fields such as...
- **The importance of** strength anisotropy has been demonstrated by...
- Convection heat transfer phenomena **play an important role in** the development of...
- For **more than 100 years** researchers have been observing the stress-strain behaviour of...
- **Much research in recent years has focused on** carbon nanotubes.

2. VERBS USED IN THE LITERATURE REVIEW TO PRESENT PREVIOUS AND/OR CURRENT RESEARCH AND CONTRIBUTIONS

achieve	develop	obtain
address	discover	overcome
adopt	discuss	perform
analyse	enhance	point out
apply	establish	predict
argue	estimate	present
assume	evaluate	produce
attempt	examine	propose
calculate	explain	prove
categorise	explore	provide
carry out	extend	publish
choose	find	put forward
claim	focus on	realise
classify	formulate	recognise
collect	generate	recommend
compare	identify	record
concentrate (on)	illustrate	report
conclude	implement	reveal
conduct	imply	revise
confirm	improve	review
consider	incorporate	show
construct	indicate	simulate
correlate	interpret	solve
deal with	introduce	state
debate	investigate	study
define	measure	support
demonstrate	model	suggest
describe	monitor	test
design	note	undertake
detect	observe	use
determine	prefer	utilise

Here are some examples of how these are used:

- This phenomenon **was demonstrated** by…
- In their study, expanded T-cells **were found** in…
- Initial attempts **focused on identifying** the cause of…
- Weather severity **has been shown to**…
- Early data **was interpreted** in the study by…
- The algorithm **has been proposed** for these applications…
- The results on pair dispersion **were reported in**…
- Their study **suggested** a possible cause for…
- An alternative approach **was developed** by…

Note: You can recycle these verbs at the end of the Introduction when you say what you plan to do in your paper (see **4** below)

3. GAP/QUESTION/PROBLEM/CRITICISM

This is often signalled by words such as however, although, while, nevertheless, despite, but.

ambiguous	(the) absence of
computationally demanding	(an) alternative approach
confused	(a) challenge
deficient	(a) defect
doubtful	(a) difficulty
expensive	(a) disadvantage
false	(a) drawback
far from perfect	(an) error
ill-defined	(a) flaw
impractical	(a) gap in our knowledge
improbable	(a) lack
inaccurate	(a) limitation
inadequate	(a) need for clarification
incapable (of)	(the) next step
incompatible (with)	no correlation (between)
incomplete	(an) obstacle
inconclusive	(a) problem
inconsistent	(a) risk
inconvenient	(a) weakness
incorrect	

ineffective	(to be) confined to
inefficient	(to) demand clarification
inferior	(to) disagree
inflexible	(to) fail to
insufficient	(to) fall short of
meaningless	(to) miscalculate
misleading	(to) misjudge
non-existent	(to) misunderstand
not addressed	(to) need to re-examine
not apparent	(to) neglect
not dealt with	(to) overlook
not repeatable	(to) remain unstudied
not studied	(to) require clarification
not sufficiently + adjective	(to) suffer (from)
not well understood	
not/no longer useful	few studies have...
of little value	it is necessary to...
over-simplistic	little evidence is available
poor	little work has been done
problematic	more work is needed
questionable	there is growing concern
redundant	there is an urgent need...
restricted	this is not the case
time-consuming	unfortunately
unanswered	
uncertain	
unclear	
uneconomic	
unfounded	
unlikely	
unnecessary	
unproven	
unrealistic	
unresolved	
unsatisfactory	
unsolved	
unsuccessful	
unsupported	

Here are some examples of how these are used:

- **Few researchers have addressed the problem** of…
- **There remains a need for** an efficient method that can…
- However, light scattering techniques have been **largely unsuccessful** to date.
- The high absorbance makes this **an impractical option** in cases where…
- **Unfortunately**, these methods do not always guarantee…
- **An alternative approach** is necessary.
- The function of these proteins **remains unclear**.
- These can be **time-consuming** and are often **technically difficult** to perform.
- **Although** this approach improves performance, it results in **an unacceptable** number of…
- Previous work has focused **only** on…
- However, the experimental configuration was **far from optimal**.

Note: Some of these words/phrases express very strong criticism. A useful exercise is to put an asterisk (*) next to those you think you could use if you were talking about the research of your professor or supervisor. You can also alter them to make them more polite (*i.e.* instead of *unsuccessful*, which is quite a strong criticism, you could write *may not always be completely successful*).

(to) attempt	(is) organised as follows:	(were/are) able to
(to) compare	(is) set out as follows:	accurate/accurately
(to) concentrate (on)	(is/are) presented in detail	effective/effectively
(to) conclude	(our) approach	efficient/efficiently
(to) describe	(the) present work	excellent results
(to) discuss	(this) paper	innovation
(to) enable	(this) project	new
(to) evaluate	(this) report	novel method
(to) expect	(this) section	powerful
	(this) study	practical

4. THE PRESENT WORK

(to) facilitate (to) illustrate (to) improve (to) manage to (to) minimise (to) offer (to) outline (to) predict (to) present (to) propose (to) provide (to) reveal (to) succeed	(this) work begin by/with close attention is paid to here overview	simple straightforward successful valuable aim goal intention objective purpose

Here are some examples of how these are used:

- **This paper focuses on…**
- **The purpose of this study is to describe and examine…**
- **In order to** investigate the biological significance…
- **In this paper we present…**
- New correlations were developed with **excellent** results…
- **In the present study** we performed…
- **This paper introduces** a scheme which solves these problems.
- **The approach we have used in this study** aims to…
- **This study** investigated the use of…
- **In this report** we test the hypothesis that…
- **This paper is organised as follows:…**

Note: In a thesis or a very long research paper, you use these to say what each chapter or section will do. Don't rely on one-size-fits-all verbs such as *discuss*; some chapters/sections do not 'discuss' anything, and even if they do, their main purpose may be to *compare* things, *analyse* things or *describe* things rather than to *discuss* them.

1.5 Writing an Introduction

In the next task, you will bring together and use all the information in this unit. You will write an Introduction according to the model, using the grammar and vocabulary you have learned, so make sure that you have the model (Section 1.3.3) and the vocabulary (Section 1.4) in front of you.

Throughout this unit you have seen that conventional science writing is easier to learn, easier to write and easier for others to read than direct translations from your own language or more creative writing strategies. You have learned the conventional model of an Introduction and collected the vocabulary conventionally used. Your sentence patterns should also be conventional; use the sentences you have read in your target articles and in the Introductions printed here as models for the sentence patterns in your writing, and adapt them for the task.

Follow the model exactly this time. After you have practiced it once or twice you can vary it to suit your needs. However, you should use it to check Introductions you have written so that you can be sure that the information is in an appropriate order and that you have done what your readers expect you to do in an Introduction.

Although a model answer is provided in the Key, you should try to have your own answer checked by a native speaker of English if possible, to make sure that you are using the vocabulary correctly.

1.5.1 Write an Introduction

Imagine that you have just completed a research project to design a bicycle cover which can protect the cyclist from injury, pollution, or just from rain. Perhaps you provided a computer simulation of its use, or modelled the ventilation system. Perhaps you were involved in the aerodynamics, or the polymer construction of the material for the cover — or any other aspect of the project. Write the Introduction of your research paper, to be published in the *Journal of Pedal-Powered Vehicles* (Vol. 3). The title of your research paper is **A COVER FOR THE SPPPV (Single-Person Pedal-Powered Vehicle)** and your Introduction should be between 200–400 words. You can lie as much as you like, and of course you will have to create fake research references. Follow the model as closely as possible;

make sure your Introduction contains the four main components of the model and try out some of the new vocabulary.

If you get stuck and don't know what to write next, use the model and the vocabulary to help you move forward. Don't look at the key until you have finished writing.

1.5.2 Key

Here is a sample answer. When you read it, think about which part of the model is represented in each sentence.

A COVER FOR THE SPPPV (Single-Person Pedal-Powered Vehicle)

Concern about global warming and urban air pollution have become central issues in transport policy decision-making, and as a result much research in recent years has focused on the development of vehicles which are environmentally friendly. Air quality in cities is currently significantly lower than in rural areas[1] and this has been shown to be directly linked to the level of vehicle emissions from private cars.[2] Due to the fact that urban transport policy in the UK is designed to reduce or discourage the use of private cars,[3] there has been an increase in the sale of non-polluting vehicles such as the SPPPV (Single-Person Pedal-Powered Vehicle). However, although the number of SPPPV users has increased, safety and comfort issues need to be addressed if the number of users is to increase to a level at which a significant effect on environmental pollution can be achieved.

Researchers have studied and improved many aspects of the SPPPV. In 1980, Wang *et al.* responded to the need for increased safety by designing an SPPPV surrounded by a 'cage' of safety bars,[4] and in 2001 Martinez developed this further with the introduction of a reinforced polymer screen which could be fitted to the safety bars to protect the cyclist's face in the event of a collision.[5] The issue of comfort has also been addressed by many design teams; in 1998 Kohl *et al.* introduced an SPPPV with a built-in umbrella, which could be opened at the touch of a button,[6] and more recently, Martinez[7] has added a mesh filter which can be placed over the entire cage to reduce the risk of environmental pollution. However, the resulting 'cage' or cover is aerodynamically ineffective due to the shape of the umbrella and the weight of the mesh filter.

In this study, we used computer simulation to model the aerodynamic effect of the existing safety and comfort features and we present a new design which integrates these features in an optimally-effective aerodynamic shape.

Unit 2 ✏ Writing about Methodology

2.1 Structure

The title of this section varies in different disciplines and in different journals. It is sometimes called *Materials and Methods,* or it can be called *Procedure, Experiments, Experimental, Simulation, Methodology* or *Model.* This section is the first part of the central 'report' section of the research article (the second part is the Results section), and it reports what you did and/or what you used.

Most journals publish (usually on the Internet) a Guide for Authors. Before you begin to read this unit, access the guide for a journal you read regularly — if you're lucky, it will include a short description of what the editors expect in each section in addition to technical information relating to the figures. Here is a typical sentence from such a guide:

> *The Methodology should contain sufficient detail for readers to replicate the work done and obtain similar results.*

It is true that your work must contain sufficient detail to be repeatable, but the type of writing you will need to do is not just a record of what you did and/or used. One of the most interesting and important changes you need to make in the way you write is that until now, you have probably been writing for people (perhaps your teachers) who know more about your research topic than you do. You have been displaying to them that you understand the tasks they have set and have performed them correctly. However, when you write a research article, people will be learning from you. Therefore you now need to be able to communicate information about a new procedure, a new method, or a new approach so that everyone reading it can not only carry it out and obtain similar results, but also understand and accept your procedure.

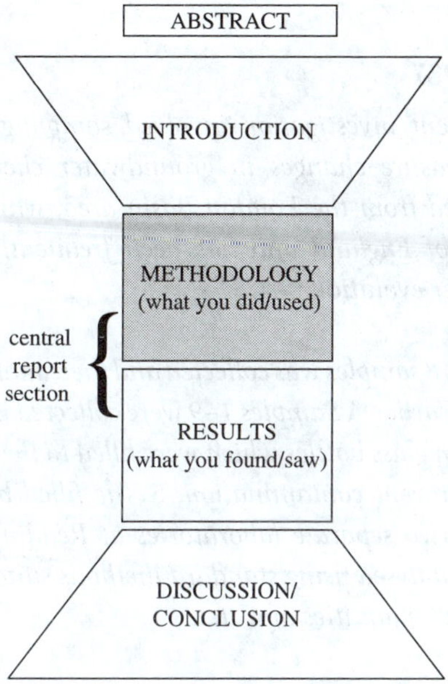

Fig. 1. The shape of a research article or thesis.

When we come to ask our three questions:

- How do I start the Methodology/Experiments section? What type of sentence should I begin with?
- What type of information should be in this section, and in what order?
- How do I end this section?

you already know that the Methodology should contain a detailed description of what you did and/or used, and this helps to answer the second of the three questions. As we will see, however, it is not a full answer; to be effective and conform to what is normally done in a research paper, this section must contain other important information as well.

Read the example below. The title of the paper is **Changes in the chemistry of groundwater in the chalk of the London Basin.** Don't worry if the subject matter is not familiar to you or if you have difficulty understanding individual words, especially technical terms like *groundwater.* Just try to get a general understanding at this stage and familiarise yourself with the type of language used.

Methodology

1 *The current investigation involved sampling and analysing six sites to measure changes in groundwater chemistry.* **2** *The sites were selected from the London Basin area, which is located in the south-east of England and has been frequently used to interpret groundwater evolution.*[2,3,4]

3 *A total of 18 samples was collected and then analysed for the isotopes mentioned earlier.* **4** *Samples 1–9 were collected in thoroughly-rinsed 25 ml brown glass bottles which were filled to the top and then sealed tightly to prevent contamination.* **5** *The filled bottles were shipped directly to two separate laboratories at Reading University, where they were analysed using standard methods suitably miniaturised to handle small quantities of water.*[5]

6 *Samples 10–18 were prepared in our laboratory using a revised version of the precipitation method established by the ISF Institute in Germany.*[6] **7** *This method obtains a precipitate through the addition of $BaCl_2.2H_2O$; the resulting precipitate can be washed and stored easily.* **8** *The samples were subsequently shipped to ISF for analysis by accelerator mass spectrometry (AMS).* **9** *All tubing used was stainless steel, and although two samples were at risk of CFC contamination as a result of brief contact with plastic, variation among samples was negligible.*

2.2 Grammar and Writing Skills

This section deals with three language areas which are important in the Methodology:

PASSIVES AND TENSE PAIRS
USE OF 'A' AND 'THE'
ADVERBS AND ADVERB LOCATION

2.2.1 Passives and tense pairs

When a sentence changes from active to passive, it looks like this:

The dog bit the policeman.	*active*
The policeman was bitten by the dog.	*passive*

But in formal academic writing, when you report what you did, you don't write 'by us' or 'by me' when changing the sentence from active to passive. You simply leave the agent out, creating an agentless passive:

We/I collected the samples.	*active*
The samples were collected.	*passive*

Before you begin to write the description of what you did and used, you need to check with the Guide for Authors in your target journal (if you are writing a doctoral thesis in an English-speaking country, check with your supervisor) to find out whether this part of the paper or thesis should be written in the passive or in the active. You can use the active (*we collected*) if you worked as part of a research team. Using the active is not usually appropriate when you write your PhD thesis because you worked alone, and research is not normally written up in the first person singular (*I collected*). In most cases, you will find that in papers and theses, the procedure you used in your research is described in the passive, either in the **Present Simple passive** (*is collected*) or in the **Past Simple passive** (*was collected*). To make that choice, it is useful to explore the advantages and disadvantages of each.

There are two common errors in the way passives are used in this section. First, look at these two sentences:

(a) *A flexible section is inserted in the pipe.*	*Present Simple passive*
(b) *A flexible section was inserted in the pipe.*	*Past Simple passive*

When you write about what you did and what you used, you need to be able to distinguish between standard procedures, *i.e.* what is normally done or how a piece of equipment is normally constructed, and what you did yourself. In the examples above, (a) uses the Present Simple tense to describe what is normally done or to describe a standard piece of equipment used in the research and (b) uses the Past Simple tense to describe what you did yourself. It is conventional in this section to use the passive for both, and the agent of the action is not mentioned in the sentence — we don't add 'by the researcher' or 'by me' at the end.

Passives used in formal writing are normally of this type, *i.e.* agentless passives. However, because the agent is not given, the only way that the reader can separate what is normally done (Sentence (a)) from what you did yourself (Sentence (b)) is if you use the correct tense. Check your target journal, but wherever possible it is clearer to use the Present Simple passive for what is normally done and the Past Simple passive to indicate what you did yourself.

You can see that if you don't pay careful attention to the tense of these sentences, your own work may become confused with the standard procedures you are describing. This is a very common error, even among native speakers, and has serious consequences. If the reader cannot identify your contribution, that is a disaster! Look at this example:

> *Two dye jets* ***are*** *placed in the laser cavity. A gain jet* ***is*** *then excited by an argon ion laser and the pulses* ***are*** *spatially filtered in order to obtain a Gaussian beam. Polarisation* ***is*** *confirmed using a polarising cube. The pulses* ***were*** *split into reference pulses and probe pulses and the reference pulses* ***were*** *carefully aligned into the detector to minimise noise levels.*

In this case, *splitting the pulses into two groups for testing* was the significant innovation of the writer's research team but the only way the reader knows this is because of the change in tense from Present Simple passive to Past Simple passive (***were*** *split*). Here is another example:

> *Samples for gas analysis* ***were collected*** *using the method described by Brown (1999), which* ***uses*** *a pneumatic air sampling pump.*

Another difficulty arises with the passive when you write about the procedure you used and compare it with the work of other researchers. You can use the Past Simple agentless passive to describe the procedure you used (*the samples* ***were collected*** *using a suction tube*) but you may also need to use exactly the same Past Simple agentless passive to describe the procedure used by the other researcher whose work you are citing (*the samples* ***were collected*** *using a suction tube*). This means that unless you are very careful, the reader has no way of separating your work from that of the other researcher. The fact that you are so familiar with what you did means that your own contribution is obvious to you — but it may not be obvious to your reader.

One way to make sure that your own contribution is clear and easy to identify is by marking it with words — perhaps by adding phrases like ***In this study***, *the samples were collected using a suction tube* or ***In our experiments*** *the samples were collected using a suction tube*, and by identifying the procedure used by other researchers with careful references at the appropriate place in the sentence (*In Brown (1999) the samples* ***were collected*** *using a suction tube*).

There are five possible uses that you may need. Note the different tenses.

	What do you mean?	How can you make it clear?
1	*X was (collected/ substituted/ adjusted etc.)* by **me** in the procedure or work that **I** carried out	Either move to the active (*We collected/adjusted/ substituted etc.*) or add words or phrases such as *here/in this work/in our model* or use a 'dummy' subject such as *This experiment/The procedure*
2	*X was (collected/ substituted/ adjusted etc.)* by the person whose procedure or work I am using as a basis for, or comparing with, my own	Give a research reference and/ or add words/phrases such as *in their work/in that model*
3	*X is (collected/substituted/ adjusted etc.)* normally, *i.e.* as part of a standard procedure	You may need a research reference even if it is a standard procedure, depending on how well-known it is. Use phrases such as *as in* [5]

4	*X is (collected/substituted/ adjusted etc.)* as you can see in Fig. 1, but it was collected/ substituted/adjusted *etc.* by me	Move to the active (*We collected/adjusted/substituted etc.*) if you can or make sure that you come out of the Present Simple passive when you stop describing the figure
5	*X is (collected/substituted/ adjusted etc.)* by me in the procedure/work that **I** carried out, but my field requires authors to write procedural descriptions in the Present Simple tense. (This is quite common in pure mathematics)	Either move to the active ('*We collect/adjust/substitute etc.*) or add words or phrases such as *here/in this work/ in our model* or use a 'dummy' subject such as *This experiment/The procedure*

2.2.2 Use of 'a' and 'the'

This is one of the most problematic areas of English grammar and usage. Many languages do not have separate words for **a** and **the**, and even if they do, these words may not correspond exactly to the way in which they are used in English. Students studying English as a second language are often given the following useful, but sometimes confusing, rule:

SINGULAR COUNTABLE NOUNS NEED A DETERMINER

A determiner is a word like **the**, **a**, **my**, **this**, **one**, **some**. It's a difficult rule to operate successfully because two problems need to be solved before you can use it. Firstly, it's hard to know exactly which nouns are countable and, secondly, even when you know, how do you decide whether to use **a** or **the**?

Let's look at the first problem. Deciding which nouns are countable nouns and which aren't isn't as easy as it looks. Many nouns which are often considered uncountable can actually be used 'countably'. Nouns like *death* or *childhood*, for example, can occur in the plural:

*There have been three **deaths** this year from pneumonia.*
*Our **childhoods** were very different; I grew up in France and she grew up in China.*

and so can nouns like industry:

Many industries rely on fossil fuels.

Even names of materials like *steel* can occur in the plural:

*Some **steels** are used in the manufacture of medical instruments.*

In the following list of uncountable nouns, mark those which can also be used in the plural, *i.e.* countably. The way you use a noun determines whether it is used in its countable or uncountable form. So when you use a noun like *industry*, stop and think — do you mean industry in general (uncountable) or a particular industry (countable)? Check your answers in the Key.

absence	access	analysis	advice	age
agriculture	cancer	art	atmosphere	beauty
behaviour	duty	capacity	childhood	calculation
concern	economy	death	democracy	depression
design	environment	earth	education	electricity
energy	evidence	equipment	existence	experience
failure	fashion	fear	fire	health
food	freedom	history	growth	independence
heat	help	insurance	ice	knowledge
industry	information	machinery	intelligence	light

life	luck	philosophy	nature	loss
paper	organisation	pollution	physics	oil
power	progress	research	protection	policy
pressure	reality	security	respect	purity
rain	sand	strength	silence	safety
salt	science	time	stuff	sleep
swimming	space	trouble	trade	sunlight
transport	technology	waste	truth	traffic
vision	treatment	water	velocity	violence
wildlife	wind	work	wealth	welfare

KEY
The nouns which can also have a countable meaning appear in italics.

absence	access	*analysis*	advice	*age*
agriculture	*cancer*	*art*	*atmosphere*	*beauty*
behaviour	*duty*	*capacity*	*childhood*	*calculation*
concern	*economy*	*death*	*democracy*	*depression*
design	*environment*	earth	*education*	electricity
energy	evidence	equipment	*existence*	*experience*
failure	*fashion*	*fear*	*fire*	health
food	freedom	*history*	*growth*	independence
heat	help	insurance	ice	knowledge
industry	information	machinery	intelligence	*light*
life	luck	*philosophy*	nature	*loss*
paper	*organisation*	pollution	physics	*oil*
power	progress	research	protection	*policy*

pressure	*reality*	security	respect	*purity*
rain	*sand*	*strength*	*silence*	safety
salt	*science*	*time*	stuff	sleep
swimming	*space*	*trouble*	*trade*	sunlight
transport	*technology*	*waste*	*truth*	traffic
vision	*treatment*	*water*	*velocity*	violence
wildlife	*wind*	work	wealth	welfare

Now look at the second problem: how do you decide whether to use **a** or **the**? You may have been told that **a** is used for general reference and **the** is used for specific reference, but in the following sentence:

*There is **a** book on the shelf above my desk; can you bring it here?*

***a** book* clearly refers to a specific book; in fact, that part of the sentence specifies which book the speaker wants. So if the specific/general criterion doesn't help you to select **a** or **the**, what does?

Start by asking yourself this simple question: Why do you use **a** the first time you talk about something, but when you refer to it again you use **the**? After all, it's the same specific item on both occasions. For example, in the sentence below, why does the first reference to the cheese sandwich use **a** and the second reference use **the** if both refer to the same specific sandwich?

*I had **a** cheese sandwich and **an** apple for lunch. **The** sandwich was fine but **the** apple had a worm in it.*

The difference is that the first time the speaker mentions the *cheese sandwich* or the *apple*, only the speaker knows about them — but the second time, both the speaker and listener know. The *worm*, however, is 'new' to the listener, and so is referred to using **a.** Now we can add a new rule:

> USE **THE** IF OR WHEN YOU AND YOUR READER BOTH KNOW WHICH THING/PERSON YOU MEAN.

This is true even if the thing or person has not been mentioned before, for example, in the following sentences:

I arrived at Heathrow Airport but ***the*** *check-in was closed.*
I bought a new computer but ***the*** *keyboard was faulty.*

check-in and *keyboard* need **the** because as soon as Heathrow Airport is mentioned, the speaker and listener know about and therefore share *check-in*; as soon as a computer is mentioned, they share *keyboard*. Similarly, in the sentence:

He lit a match but the flame went out.

mentioning *a match* automatically creates the concept of *flame* in the reader's mind — and this shared understanding is marked by the use of **the**. Similarly, if we were in the same room and I told you to look up at **the** ceiling, you wouldn't ask me 'Which ceiling are you talking about?' because it would be obvious; we would share it.

Did she get ***the*** *job?* (the job we both know she wanted)
I'll meet you in ***the*** *library later.* (the library we normally use)

Here are some more useful rules:

USE **THE** IF THERE IS ONLY ONE POSSIBLE REFERENT

We removed ***the*** *softest layer of membrane.*
Cairo is ***the*** *capital of Egypt.*
The opening was located in ***the*** *centre of each mesh.*
Government policy is committed to protecting ***the*** *environment.*
The *sun's altitude is used to determine latitude.*

USE **A** IF IT DOESN'T MATTER *or* YOU DON'T KNOW *or* YOUR READER DOESN'T KNOW WHICH THING/PERSON YOU ARE REFERRING TO.

A 35 ml brown glass bottle was used to store the liquid. (It doesn't matter which 35 ml brown glass bottle was used.)

The subject then spoke to an interviewer. (It doesn't matter which interviewer/I know which one but you don't.)

It works on the same principle as a combustion engine. (It doesn't matter which combustion engine.)

Sometimes the choice of **a** or **the** changes the meaning of the sentence completely:

(a) *This effect may hide* ***a*** *connection between the two.* (There may possibly be a connection between the two but if there is, we cannot see it.)
(b) *This effect may hide* ***the*** *connection between the two.* (There is definitely a connection between the two but we may not be able to see it because of *this effect*.)

Here's another pair in which the choice of **a** or **the** has a significant effect on the meaning (∅ is used here to indicate the plural of **a**):

(a) *The nodes should be attached to ∅ two adjacent receptor sites.* (There are many receptor sites and any two adjacent ones will do.)
(b) *The nodes should be attached to* ***the*** *two adjacent receptor sites.* (There are only two receptor sites.)

The best way to use the information you have just learned is to take a paragraph from a research article that you are reading and use the information in this grammar section to work out why the writer has chosen each instance of **the** or **a,** or why the writer has not used any determiner before a particular noun.

Another important point to note about the use of **a**, **the** and ∅ is that they can all be used generically, *i.e.* when expressing a general truth:

The electroencephalograph is a machine for measuring brain waves.
An electroencephalograph is a machine for measuring brain waves.
Electroencephalographs are machines for measuring brain waves.

One last note: **a** is used before consonant sounds, while **an** is used before vowel sounds. Sound, not spelling, is important here, so we write ***an*** *MRI scan* because the letter 'M' is pronounced 'em', but ***a*** *UV light* because the letter 'U' is pronounced 'yoo'.

2.2.3 Adverbs and adverb location

When you are communicating complex ideas in another language, an obvious grammatical error is not as bad as an error which is invisible. A proofreader or editor will notice an obvious grammatical error and correct it, but if the sentence is written in grammatically correct English the error is not visible to proofreaders and editors. An example of an invisible error is where the sentence is grammatically correct but the choice of which verb tense to use is inappropriate or does not represent the intention of the writer. These hidden errors are worrying because neither the writer nor the editor/proofreader knows they have occurred and yet the sentence does not mean what the writer intended.

Common hidden errors include mistakes in the use of **a** and **the** (see Section 2.2.2 above), whether or not to use a comma before the word *which* in relative clauses and adverb location errors. Adverb location errors are easy to make and hard to detect.

Adverbs don't always do what you want or expect them to do. In the first place, adverbs needing prepositions can be ambiguous (*Look at that dog* ***with one eye*** can either mean *USING one eye* or *HAVING one eye*) and in the second place, adverbs may attach themselves to unexpected parts of a sentence. Be careful where you put your adverb, and be especially careful if you are using more than one adverb in a sentence. Here is an example of the kind of problem you may encounter:

The patient was discharged from hospital after being shot in the back with a 9 mm gun.

Did the doctors shoot her?

He gave a lecture about liver cancer at the hospital last January.

Was the lecture in the hospital — or the cancer? Did the lecture refer to cancer cases occurring in January or did the lecture itself occur in January?

Although there are rules for adverb location, they are complex and hard to apply when you are writing. Since your aim is to stay safe and write clearly, it is better to avoid adverb clusters like these, and rewrite the information in a different order. If your adverb relates to the whole sentence (*i.e. clearly, last January, as a result)* then consider putting the adverb at the front of the sentence:

Last January he gave a lecture about liver cancer at the hospital

If you are still left with ambiguous adverb clusters, consider breaking the sentence down into units, each with its own adverb:

Last January he gave a lecture at the hospital; his subject was liver cancer

2.3 Writing Task: Build a Model

2.3.1 Building a model

You are now ready to begin to build a model of the Methodology by writing a short description of what the writer is doing in each sentence in the space provided below. The Key is on the next page. Once you have tried to produce your own model, you can use the Key to help you write this section of a research article when you eventually do it on your own.

GUIDELINES

You should spend 30–45 minutes on this task.If you can't think of a good description of the first sentence, choose an easier one, for example Sentence 4, and start with that. Remember that your model is only useful if it can be transferred to other Methodology sections, so don't include content words such as *groundwater* or you won't be able to use your model to generate Methodology sections in your field.

One way to find out what the writer is doing in a sentence — rather than what s/he is saying — is to imagine that your computer has accidentally deleted it. What is different for you (as a reader) when it disappears? If you press another key on the computer and the sentence comes back, how does that affect the way you respond to the information?

Another way to figure out what the writer is doing in a sentence — rather than what s/he is saying — is to look at the grammar and vocabulary clues. What is the tense of the main verb? What is that tense normally used for? Is it the same tense as in the previous sentence? If not, why has the writer changed the tense? What words has the writer chosen to use?

Don't expect to produce a perfect model. You will modify your model when you look at the Key, and perhaps again when you compare it to the way Methodology sections in your target articles work.

Changes in the chemistry of groundwater in the chalk of the London Basin	
Methodology	**In this sentence, the writer:**
1 *The current investigation involved sampling and analysing six sites to measure changes in groundwater chemistry.* **2** *The sites were selected from the London Basin area, which is located in the south-east of England and has been frequently used to interpret groundwater evolution.*[2, 3, 4]	1__________ 2__________
3 *A total of 18 samples was collected and then analysed for the isotopes mentioned earlier.* **4** *Samples 1–9 were collected in thoroughly-rinsed 25 ml brown glass bottles which were filled to the top and then sealed tightly to prevent contamination.* **5** *The filled bottles were shipped directly to two separate laboratories at Reading University, where they were analysed using standard methods suitably miniaturised to handle small quantities of water.*[5]	3__________ 4__________ 5__________
6 *Samples 10–18 were prepared in our laboratory using a revised version of the precipitation method established by the ISF Institute in Germany.*[6] **7** *This method obtains a precipitate through the addition of $BaCl_2.2H_2O$; the resulting precipitate can be washed and stored easily.*	6__________ 7__________

8 *The samples were subsequently shipped to ISF for analysis by accelerator mass spectrometry (AMS).* **9** *All tubing used was stainless steel, and although two samples were at risk of CFC contamination as a result of brief contact with plastic, variation among samples was negligible.*	8________________ 9_________________

2.3.2 Key

In Sentence 1 *'The current investigation involved sampling and analysing six sites to measure changes in groundwater chemistry.'* the writer **offers a general overview of the entire subsection, including the purpose of the investigation.**

If you wrote 'introduction' or 'introduces the Methodology' here, that won't help you when you come to write your own thesis or research article because it doesn't tell you what exactly to write in that sentence.

Why do I need to introduce the Methodology?

In some cases, writers begin immediately with a description of the procedure or the materials. This is appropriate where the research focus is very narrow and all those who are likely to read it are carrying out similar research. If this is not the case, it is more reader-friendly to start with some introductory material. The aim of providing a short introduction is to make the entry to that section smooth for the reader. There are many ways to introduce the Methodology. Here are three of the most common ways:

- Offer a general overview by outlining the parameters of the work, for example the number of tests, the equipment /material/software used and perhaps also the purpose of the investigation. This helps the reader to get a general idea of this section.
- Provide background information about the materials or about the source of the materials/equipment.
- Refer back to something in the previous section. Common options are restating the aim of the project or the problem you are hoping to address.

If you start with a general overview or even a general paragraph about what was done and used, it can then be broken down to produce the details. However, if you begin with the details, you force the reader to put those details together to create a general picture of what you did and used. This is quite difficult for the reader to do and it is not his/her job; it is your job as a writer to arrange the information in an appropriate order so that it is easy for the reader to process it.

Furthermore, asking your reader to put details together to create a picture of what you did is risky, because each reader may create a slightly different picture of the process if they begin 'bottom-up' with the details, rather than 'top-down' with a general overview. When you write using 'top-down' strategies you are in control. If you begin with general statements about what was done/used (*In all cases, Most sites*), you and your reader share the same framework, so when you fill in the details you are creating the same picture of what was done/used in the mind of each individual reader. Remember: show your reader the wall before you begin to examine the bricks.

In Sentence 2 '*The sites were selected from the London Basin area, which is located in the south-east of England and has been frequently used to interpret groundwater evolution.*[2–4] the writer provides background information and justifies the choice of location by referring to previous research.

Why do I need to justify or give reasons for what I did? Isn't it obvious?

Your reasons may be obvious to you, but they are not always obvious to your readers. If you fail to provide justification for what you did, then the reader may not accept the validity of your choices. They may wonder why you did things in a particular way, or why you used a particular procedure. This has a negative effect: if you don't explain why you did things then readers cannot be expected to accept your methodology, and this will eventually affect the way they evaluate your whole paper.

Many writers believe that this section is just an impersonal description of what was done or used; in fact there is a strong persuasive and communicative element. We see this not only in language such as *thoroughly* or *with care* but also in the frequency of justification. In this description of your materials and methods, you need to communicate not only **This is exactly what I did/used** but also **I had good reasons for those decisions.** Justification enables the reader to trust the choices you made.

Sometimes background information is given in the Present Simple to justify choices made. For example, you may have chosen a particular material because of its properties; if so, say what those properties are (*This material is able to...*). You may have chosen specific equipment or software because of what it can do; if so, say what that is. In Sentence 2, we understand that the writer chose this geographical area because it had been previously validated as an appropriate location by other researchers.

> **In Sentence 3** '*A total of 18 samples was collected and then analysed for the isotopes mentioned earlier*' **the writer provides an overview of the procedure/method itself.**

If I gave a general overview at the start of this subsection, why should I also give an overview of the procedure itself?

As you saw in Section 1.2.4, the beginning of a paragraph often signals the beginning of a new topic, and providing an introductory sentence is a reader-friendly technique. In addition, the overview in Sentence 3, like the one at the start of the subsection, enables the writer to move in a 'top-down' direction by creating a general framework into which the details can be easily slotted. Because the reader knows from the start how many samples were tested and what was done with them, both reader and writer share the same clear picture. These sentences often start with phrases like *Most of the tests* or *In all cases* (see the vocabulary list in Section 2.4.2).

> **In Sentence 4** '*Samples 1–9 were collected in thoroughly-rinsed 25 ml brown glass bottles which were filled to the top and then sealed tightly to prevent contamination.*' **the writer provides details about what was done and used and also shows that care was taken.**

How much detail do I need to provide?

If you're not certain that all readers are familiar with the precise details of your methodology, it is better to give slightly too much information than too little. By the time you write up your research you will probably have repeated your experiments or simulations many times and so you are very familiar with the materials, quantities, equipment, software, the sequence or steps in the procedure and the time taken for each step. Because of this familiarity, specific details (the size of the bottles in Sentence 5, for example) may seem obvious to you, but those details may not be obvious to every reader. If you want another researcher to be able to reproduce your work and obtain similar results, you should include every specification and detail.

Note that in this sentence, the writer uses *thoroughly*, *filled to the top* and *tightly* to communicate to the reader that the work was carried out with care. Remember that your aim in writing the paper is not only to say what you did and found, but also to make sure that your reader accepts the conclusions at the end of your paper. In order to do this, the reader has to accept your results — but to accept your results s/he must first accept your methodology. For this reason, it is important to present yourself as a competent researcher who carries out procedures accurately and with care.

Notice the use of *25 ml* in Sentence 4. *ml* is the SI (Système International d'Unités) symbol for *millilitre*. Check the SI to make sure that you are using the correct symbol. There is often a space between the quantity/number and the SI symbol; in addition, although SI symbols look like abbreviations they are not, and therefore should not be followed by a period.

> **In Sentence 5 '*The filled bottles were shipped directly to two separate laboratories at Reading University, where they were analysed using standard methods suitably miniaturised to handle small quantities of water.*[5]' the writer continues to describe what was done in detail, using language which communicates that care was taken.**

Can you see which words in Sentence 5 communicate to the reader that care was taken? The writer could just have written *The filled bottles were shipped to two laboratories and analysed using standard methods miniaturised to handle small quantities of water*, but including words like *directly, separate* and *suitably* communicates reliability.

In Sentence 6 '*Samples 10–18 were prepared in our laboratory using a revised version of the precipitation method established by the ISF Institute in Germany.*[6]' **the writer describes what was done by referring to existing methods in the literature.**

Why should I refer to other research; why not just describe the method I used?

One reason is that it is unlikely that you created the entire method you used all by yourself. In many cases part of it will be taken from a method used or discovered by someone else and their method may be very well known, so if you give the research reference you do not need to give every detail. Giving the research reference, therefore, provides you with a shortcut. You will find vocabulary for this in **Option 1** in Section 2.4.

But if the reference is available in the literature, why does the writer need to give any details? Why can't readers just go to the library, find the reference and read it themselves?

In this case, the writer provides basic details of the method because some readers may not be familiar with it and it is not always appropriate to send readers to the library or Internet to look up a reference. It's a matter of professional courtesy for writers to describe the procedures, tests, equipment or materials they used even when they are used in a way that is identical to the reference. Remember to use the **Present Simple** for this kind of background information (*This method obtains*) and to switch back to the **Past Simple** when you return to describing what you did.

Comparisons between your materials and methods and those of other researchers in the same field are a legitimate topic for the Methodology section. It is common to keep previous or current research procedures clearly in your readers' view so that they can see how your work is different from other work in the area. Either your method is identical to others you mention (**Option 1** in the vocabulary list in Section 2.4), or it is similar (**Option 2** in the vocabulary list), or it is significantly different, in which case the differences between your materials/method and those of other researchers in the same field may even represent the actual contribution of your paper/thesis itself (**Option 3**).

When you refer to the work of other researchers, be careful about the location of your reference notation in the sentence; you may accidentally credit someone with work they have not done — perhaps even with your own work! Remember that reference notations do not automatically go at the end of a sentence.

It is sometimes appropriate or necessary to mention the effects of the procedures you used. However, it is not a good idea to discuss them or comment at this stage. If you go into too much detail you may leave yourself with nothing to write about in the Results section. Interestingly, it is common to provide further details about the methodology in the Results section. Sometimes the Methodology section just provides basic parameters and the method itself is detailed in the Results section in relation to the results obtained.

In Sentence 7 '*This method obtains a precipitate through the addition of $BaCl_2 2H_2O$; the resulting precipitate can be washed and stored easily.*' the writer provides more detailed information about the method and shows it to have been a good choice.

Justification is common throughout this section; as before, the aims are to answer possible criticisms or doubts about your choices, to assure the reader that your choices were made on the basis of good reasons and to give those reasons. We often see justification of significant choices and the reason for rejecting alternative options given in full. As mentioned earlier, this is because it is essential that your reader accepts the decisions you made about your methodology.

In Sentence 8 '*The samples were subsequently shipped to ISF for analysis by accelerator mass spectrometry (AMS).*' the writer provides more details of the method.

It is interesting to note that, as mentioned earlier, you need to do more than just provide details of what you did and used; this is the only sentence in this section that gives details and nothing more — every other sentence has an additional function.

> **In Sentence 9** '*All tubing used was stainless steel, and although two samples were at risk of CFC contamination as a result of brief contact with plastic, variation among samples was negligible.*' **the writer mentions a possible difficulty in the methodology.**

Doesn't this discuss a result of what was done?

No, it's actually saying that the problems in the methodology didn't affect the results. Sometimes you do need to mention results in this section, but only if the preliminary results were used to modify or develop the design of the main experiments/simulations.

Why should I mention problems in the methodology? Won't it make me look bad?

In fact the opposite is true. In the first place, if you don't mention the imperfections in your work, it may look as though you are not aware of them, which gives a very poor impression. So you look far more professional if you **do** mention them. If you ignore or try to hide imperfections (such as a data set which was too small, equipment or software that was not ideal) and your readers notice them, they will begin to doubt your legitimacy as a researcher, which affects their acceptance of your results and conclusions.

Second, whenever you finish a piece of research, there is a good chance that you have learned enough from the problems encountered during the project to do it better next time. Should you delay writing it up while you repeat the work and improve your technique? What if you learn more this time too; should you delay again while you do it again? And again? If you do, you may never actually write it up. An acceptable option is to write up the research and acknowledge the problems or difficulties you encountered. In fact, it's not only considered acceptable to mention them in this section, it's much better to do it here rather than wait until the end. It isn't considered appropriate to mention limitations or imperfections for the first time when you are discussing suggestions for future work in the Discussion/Conclusion.

But how can I talk about problems in my work without looking like a failure?

Use vocabulary that **minimises the problem**, **minimises your responsibility**, **maximises the good aspects** and **suggests a solution**. In the example above, the writer has acknowledged that there was a problem and then minimised its effects (*variation among samples was negligible*). This is a standard way of dealing with the need to talk about problems. You can find examples of the language needed to refer to problems and difficulties in a conventional, professional way in the vocabulary list in Section 2.4.

2.3.3 The model

Here are the sentence descriptions we have collected:

In Sentence 1 the writer offers a general overview of the subsection.
In Sentence 2 the writer provides background information and justification.
In Sentence 3 the writer provides an overview of the procedure/ method itself.
In Sentence 4 the writer provides details about what was done and used and shows that care was taken.
In Sentence 5 the writer continues to describe what was done in detail, using language which communicates that care was taken.
In Sentence 6 the writer describes what was done by referring to existing methods in the literature.
In Sentence 7 the writer provides more detailed information about the method and shows it to have been a good choice.
In Sentence 8 the writer provides more details of the method.
In Sentence 9 the writer mentions a possible difficulty in the methodology.

We can streamline these so that our model has FOUR basic components. Unlike the Introduction model, in which all the items of each component are likely to be used, this is a 'menu' from which you select items appropriate to your research topic and the journal you are submitting to. If you constructed the equipment yourself you won't need to 'give the source of' the equipment used in component 1. If there were no problems, you won't need the fourth component at all.

1	PROVIDE A GENERAL INTRODUCTION AND OVERVIEW OF THE MATERIALS/METHODS RESTATE THE PURPOSE OF THE WORK GIVE THE SOURCE OF MATERIALS/EQUIPMENT USED SUPPLY ESSENTIAL BACKGROUND INFORMATION
2	PROVIDE SPECIFIC AND PRECISE DETAILS ABOUT MATERIALS AND METHODS (*i.e.* quantities, temperatures, duration, sequence, conditions, locations, sizes) JUSTIFY CHOICES MADE INDICATE THAT APPROPRIATE CARE WAS TAKEN
3	RELATE MATERIALS/METHODS TO OTHER STUDIES
4	INDICATE WHERE PROBLEMS OCCURRED

2.3.4 Testing the model

The next step is to look at the way this model works in a real Materials/ Methods section (remember it may not be called Materials and Methods) and in the target articles you have selected. Here are some full-length Methodology sections from real research articles. Read them through, and mark the model components (1, 2, 3 or 4) wherever you think you see them. For example, if you think the first sentence corresponds to number 1 in the model, write 1 next to it, *etc.*

Effects of H_2O on structure of acid-catalysed SiO_2 sol-gel films

Experimental procedure

Equal volumes of tetraethylorthosilicate (TEOS) and ethanol were mixed and stirred vigorously for 10 min at room temperature.

Then 0.1 M HCl was gradually added to the solutions, until a water to TEOS molar ratio of $R = 2$ was attained. Additional deionised water was added to give solutions with $R = 3$, 4 and 5, so that for all solutions the molecular ration TEOS:HCl was maintained, as summarised in Table 1. The solutions were placed in the refluxing bath immediately after mixing, and the temperature of the bath was increased to 70°C in 15 min, while stirring, and kept there for 2 h. The solutions were then aged for 24 h at room temperature, before being diluted with an equal volume of EtOH and stirred for 10 min, to give the solution used for spin coating. All the chemicals were obtained from Aldrich Chemicals Ltd.

The sols were dispensed on p-type, 75 mm diameter silicon wafers, through a 0.1 μm filter (PTFE Whatman, obtained from BDH Merk Ltd), and thereafter the substrate was spun at 2000 rpm for 15 s. The coated substrate was baked at 100°C for 5 min, and then cleaved into 10 pieces. Each piece was baked in air at a different temperature, in the range from 100 to 1000°C, for 30 min. The samples were kept in covered petri dishes for a few days in room conditions before the experiments were continued; this allows the completion of surface hydroxylation, and gave reproducible ellipsometer results when water is used as an adsorbate.

The thickness and refractive index of the samples were measured using a Rudolph AutoEl III ellipsometer, with an operating wavelength of 633 nm, and precisions of about ±0.002 and ±3 Å in index and thickness, respectively. For microporous films, the measured index is strongly dependent on relative humidity, because of condensation of water in the pores. By measuring the dependence of index on humidity, information about porosity can be obtained. We have extended this technique to the use of different adsorbate species, in order to probe pore sizes [3]; this, for the sake of brevity, we call molecular probe ellipsometry. In this technique, the film is placed in a sealed chamber on the sample stage of the ellipsometer; first dry N_2 gas is passed through the chamber to empty the pores of any condensed adsorbate, and then N_2 having been bubbled through the liquid adsorbate is passed over the sample to fill the pores; in each case the refractive index is measured. By assuming that all the accessible pores in dry and saturated atmospheres are completely empty or filled with adsorbate, respectively, the pore volume and index of the solid skeleton can be determined by an extension of the Lorentz-Lorenz relation [8] where n_f, n_s and n_p are the refractive indices of the film, solid skeleton and pores, respectively, and v_p is the volume fraction porosity. Measurement of n_f for both the dry and saturated films allows both v_p and n_s to be determined with the assumption that n_p has the same value as that of the bulk adsorbate in the saturated case, and of air ($n_p = 1$) in the dry case.

In order to empty the pores, an initial high flow rate of N_2 was used for a few minutes and the rate was then reduced to 1000 sccm (standard c.c per minute) for 15 min. the flow rate was kept at 100 sccm for 15 min to fill the pores. The low flow rate in this case reduces the likelihood of cooling of the sample surface, which could cause condensation on the external film surface. Comparison of the measured film thickness for wet and dry atmospheres indicated that this did not occur. The temperature inside the chamber was monitored by a thermocouple to ensure that there was no drift or alteration due to gas flow. In each case, the measurement was recorded once repeatable readings were obtained. The adsorbates used are listed in Table 2. Their average diameters were estimated using a combination of bond length data [9] and Van der Waals atomic radii [10]. All were obtained from Aldrich Chemical Ltd, except $C_{24}H_{44}O_8$ obtained from Fluka Chemie AG.

The optical quality of the films was first studied qualitatively by visual examination, and by optical microscopy. The homogeneity of the films was then investigated quantitatively by measuring the intensity of scattered light resulting from oblique reflection of a laser beam from the film-coated silicon substrate. A helium-neon laser beam, having a wavelength of 633 nm, was directed onto the sample, through a chopping wheel, at an angle 59° from the normal. The specularly reflected beam was absorbed onto a black card, and the scattered light was collected at normal incidence to the sample using a ×10 microscope objective, and measured using a silicon photodiode and a lock-in amplifier. The position of lens and angle of incidence were fixed during measurements.

The film stress, σ_f, can be determined by measuring the resulting substrate curvature [11], according to Stoney's formula:

$$\sigma_f = (E_s t^2_s / 6(1 - \nu_s) t_f)(1/r_s - 1/r_f), \quad (2)$$

where r_s and r_f are the radii of curvature of the bare substrate and substrate with film, respectively; E_s, t_s and ν_s are the Young's modulus, thickness and Poisson's ratio of the silicon substrate, respectively, and t_f is the thickness of the film. Tensile stresses are positive and compressive stresses negative; thus, a positive radius of curvature denotes a convex film surface. Entire 75 mm diameter wafers were used, and curvature was measured from plots of surface profile along 30 mm lines over the central part of the film surface using a Dektak IIA auto-levelling profilometer. To reduce inaccuracy caused by lack of axial symmetry in the wafer curvature, two scans were made, in orthogonal directions, for each measurement, and the inverse radii thus obtained were averaged. Care was taken not to use wafers which had a substantially asymmetric curvature before deposition. Wafer thicknesses, measured with a micrometer, were 390 ± 3 μm. Final film thicknesses were measured by ellipsometry and checked by patterned etching and profilometry, and interim thicknesses were estimated by interpolation. Equivalent single-layer thickness measurements indicate that the assumption that final thickness is proportional to number of layers is sufficiently accurate. For $E_s/(1 - \nu_s)$, the value 180 GPa was used [11].

In order to give an indication of the effect of water content on stress, 10 layers were deposited for each R value, using 10 s rapid thermal annealing at 1000°C in all cases.

Infrared imaging of defects heated by a sonic pulse

ii) Experiment

Our experimental setup is shown in Fig. 1. The source of the sonic excitation is a Branson, Model 900 MA 20 kHz ultrasonic welding generator, with a Model GK-5 hand-held gun. The source has a maximum power of 1 kW, and is triggered to provide a short (typically 50–200 ms duration) output pulse to the gun. The gun contains a piezoelectric transducer that couples to the specimen through the 1.3-cm-diam tip of a steel horn. In the laboratory setup, as can be seen in Fig. 1, we use a mechanical fixture to hold the sonic horn firmly against the sample surface. This setup uses a machine slide to provide reproducible alignment of the horn. Typically, a piece of soft Cu sheet is placed between the tip of the horn and the specimen to provide good sound transmission. The location of the source on the sample is chosen primarily for convenience of geometrical alignment, and since it has minimal effect on the resulting sonic IR images, typically is not changed during the course of the inspection. Sound waves at frequencies of 20 kHz in metals such as aluminium or steel have wavelengths on the order of tens of centimetres, and propagate with appreciable amplitude over distances much longer than a wavelength. For typical complex-shaped industrial parts (see, for example, the aluminium automotive part shown in Fig. 1), reflections from various boundaries of the specimen introduce countless conversions among the vibrational modes, leading to a very complicated pattern of sound within the specimen during the time that the pulse is applied. Since the speed of sound in solids is typically on the order of a few km/s, this sound field completely insonifies the regions under inspection during the time that the excitation pulse is applied. If a subsurface interface is present, say a fatigue crack in a metal, or a delamination in a composite structure, the opposing surfaces at the interface will be caused to move by the various sound modes present there. The complexity of the sound is such that relative motion of these surfaces will ordinarily have components both in the plane of the crack and normal to it. Thus, the surfaces will 'rub' and 'slap' against one another, with a concomitant local dissipation of mechanical energy. This energy dissipation causes a temperature rise, which propagates in the material through thermal diffusion. We monitor this dissipation through its effect on the surface temperature distribution. The resolution of the resulting images depends on the depth of the dissipative source as well as on the time at which the imaging is carried out.

The IR camera that we used in the setup that is shown in Fig. 1 is a Raytheon Radiance HS that contains a 256×256 InSb focal plane array, and operates in the 3–5 μm spectral region. It is sensitive (with a 1 ms integration time) to surface temperature changes of ~0.03°C, and can be operated at full frame rates up to 140 Hz with that sensitivity. We have also observed the effects reported here with a considerably less expensive, uncooled, microbolometer focal plane array camera, operating in the long wavelength (7–10 μm) of the IR.

The height of biomolecules measured with the atomic force microscope depends on electrostatic interactions

MATERIALS AND METHODS

Biological samples

Aquaporin-1 (AQP1) from human erythrocyte solubilized in octyl-f3-glucopyranoside was reconstituted in the presence of Escherichia coli phospholipids to form two-dimensional (2D) crystalline sheets (Walz *et al.*, 1994). The 2D crystals were prepared at a concentration of -0.5 mg protein/ml and 0.25 mg/ml lipid in 0.25 M NaCl, 20 mM $MgCl_2$, 20 mM 2-(N-morpholino) ethanesulfonic acid (MES) (pH 6).

Hexagonally packed intermediate (HPI) layer from Deinococcus radiodurans, a kind gift of Dr. W. Baumeister, was extracted from whole cells (strain SARK) with lithium dodecyl sulfate, and purified on a Percoll density gradient (Baumeister *et al.*, 1982). A stock solution (1 mg/ml protein) was stored in distilled water at 4°C.

Purple membranes of Halobacterium salinarium strain ET1001 were isolated as described by Oesterhelt and Stoeckenius (1974). The membranes were frozen and stored at −70°C. After thawing, stock solutions (10 mg protein/ml) were kept in distilled water at 4°C.

Porin OmpF trimers from E. coli strain BZ 1 10/PMY222 (Hoenger *et al.*, 1993) solubilized in octyl-polyoxyethylene were mixed with solubilised dimyristoyl phosphatidylcholine (99% purity;

Sigma Chemical Co., St. Louis, MO) at a lipid-to-protein ratio (w/w) of 0.2 and a protein concentration of 1 mg/ml. The mixture was reconstituted as previously described (Hoenger *et al.*, 1993) in a temperature-controlled dialysis device (Jap *et al.*, 1992). The dialysis buffer was 20 mM HEPES, pH 7.4, 100 mM NaCl, 20 mM $MgCl_2$, 0.2 mM dithiothreitol, 3 mM azide.

1,2-Dipalmitoyl-phosphatidylethanolamine (DPPE) from Sigma was solubilized in chloroform:hexane (1:1) to a concentration of 1 mg/ml. The resulting solution was diluted in buffer solution (150 mM KCl, 10 mM Tris, pH 8.4) to a concentration of 100 μg/ml.

Layered crystals

$MoTe_2$, a layered crystal of the family of transition metal dichalcogenides (Wilson and Yoffe, 1969), was employed to calibrate the piezo scanner of the AFM. It was prepared by chemical vapor transport (CVT), with chlorine or bromine as carrier gases in a temperature gradient of 100°C across the quartz ampule (Jungblut *et al.*, 1992), and was a kind gift of Y. Tomm.

Muscovite mica (Mica New York Corp., New York) was used as the solid support for all samples. Mica minerals are characterized by their layered crystal structure, and show a perfect basal cleavage that provides atomically flat surfaces over several hundreds of square microns. Their hydrophilicity and relative chemical inertness (Bailey, 1984) make them suitable for the adsorption of biological macromolecules.

Atomic force microscopy

A commercial AFM (Nanoscope III; Digital Instruments, Santa Barbara, CA), equipped with a 120-μm scanner (j-scanner) and a liquid cell, was used. Before use, the liquid cell was cleaned with normal dish cleaner, gently rinsed with ultrapure water, sonicated in ethanol (50 kHz), and sonicated in ultrapure water (50 kHz). Mica was punched to a diameter of –5 mm and glued with water-insoluble epoxy glue (Araldit; Ciba Geigy AG, Basel, Switzerland) onto a Teflon disc. Its diameter of 25 mm was slightly larger than the diameter of the supporting steel disc. The steel disc was required to magnetically mount the sample on to the piezoelectric scanner.

Imaging was performed in the error signal mode, acquiring the deflection and height signal simultaneously. The deflection signal was minimized by optimizing gains and scan speed. The height images presented were recorded in the contact mode. The scan speed was roughly linear to the scan size, at 4–8 lines/s for lower magnifications (frame size 1–25 μm). The applied force was corrected manually to compensate for thermal drift. To achieve reproducible forces, cantilevers were selected from a restricted area of one wafer. The dimensions of one tip were measured in a scanning electron microscope to calculate the mechanical properties of the cantilever (Butt *et al.*, 1993). The 120-μm-long cantilevers purchased from Olympus Ltd. (Tokyo, Japan) had a force constant of $k = 0.1$ N/m, and the 200-μm-long cantilevers purchased from Digital Instruments had a force constant of 0.15 N/m. All cantilevers used had oxide-sharpened Si3N4 tips.

Sample preparation
To minimize contamination of surfaces during exposure to ambient air, sample supports were prepared immediately before use. All buffers were made with ultrapure water (–18 MDcm^{-1}; Branstead, Boston, MA). This water contains fewer hydrocarbons than conventional bidistilled water and fewer macroscopic contaminants, both of which can influence the imaging process. Chemicals were grade p.a. and purchased from Sigma Chemie AG (Buchs, Switzerland). The buffers used were Tris-(hydroxymethyl)-aminomethane (from pH 10.2 to pH 7.2), MES (from pH 6.5 to pH 5.5), and citric acid (from pH 5.4 to pH 3.0). Macromolecular samples were checked before use by conventional negative stain electron microscopy (Bremer *et al.*, 1992) and/or by sodium dodecyl sulfate-gel electrophoresis.

The samples were diluted to a concentration of 5–10 μg/ml in buffer solution (pH 8.2, 20 mM Tris-HCl, 2100 mM; monovalent electrolyte; except for DPPE, which was not further diluted) before adsorption to freshly cleaved mica. After an adsorption time of 10–60 min, the samples were gently washed with the measuring buffer to remove weakly attached membranes. This allowed height measurements at low electrolyte concentrations, at which samples adsorb sparsely to mica (Muller *et al.*, 1997a and 1997b). Experiments requiring constant pH were performed at pH 8.2. The isoelectric points of bacteriorhodopsin, AQP1, DPPE, and OmpF are 5.2 (Ross *et al.*, 1989), 6.95 (calculated), ~10 (Tatulian, 1993), and 4.64 (calculated), respectively. Thus, at this pH, all samples had a net negative charge, except for DPPE, which had a net positive charge.

Now do the same in your target articles. We hope you obtain good confirmation of the model and can now answer the questions in Section 2.1:

- How do I start this section? What type of sentence should I begin with?
- What type of information should be in this section, and in what order?
- How do I end this section?

2.4 Vocabulary

In order to complete the information you need to write this section of your paper you now need to find appropriate vocabulary for each part of the model. The vocabulary in this section is taken from over 600 research articles in different fields, all of which were written by native speakers and published in science journals. Only words/phrases which appear frequently have been included; this means that the vocabulary lists contain words and phrases which are considered normal and acceptable by both writers and editors.

In the next section we will look at vocabulary for the following seven areas of the model:

1. PROVIDE A GENERAL INTRODUCTION AND OVERVIEW OF THE MATERIALS/METHODS and GIVE THE SOURCE OF MATERIALS/EQUIPMENT USED

This includes phrases such as *In this study, most of the samples were tested using a...* as well as verbs such as *were supplied by*. A good list of commonly-used words and expressions will encourage you to include this in your first sentences.

2. SUPPLY ESSENTIAL BACKGROUND INFORMATION

This list provides words and phrases used to describe instruments, equipment or locations, and includes items such as *parallel to* and *equidistant*. They are essential because the reader needs them in order to visualise or recreate your work.

3. PROVIDE SPECIFIC AND PRECISE DETAILS ABOUT MATERIALS AND METHODS (*i.e.* quantities, temperatures, duration, sequence, conditions, locations, sizes)

This includes verbs which specifically describe what you did/used. Instead of writing only *was done* or *was used*, a more specific verb such as *optimise* or *extract* can save you time by explaining exactly what was 'done'.

4. JUSTIFY CHOICES MADE

This includes phrases that introduce the reasons for the choices you made, such as *in order to*. It also includes a list of verbs that specify the advantages of the choices you made, like *enable* and *facilitate*.

5. INDICATE THAT APPROPRIATE CARE WAS TAKEN

This includes adjectives (*careful*) as well as adverbs (*carefully*), so as to give you maximum flexibility when you are constructing sentences.

6. RELATE MATERIALS/METHODS TO OTHER STUDIES

This provides you with ways to distinguish between procedures/materials/tests which were **exactly the same** as those used by other researchers, procedures/materials/tests which were **similar to** those used by other researchers and procedures/materials/tests which were **significantly different**.

7. INDICATE WHERE PROBLEMS OCCURRED

This list includes ways of minimising the problem, minimising your responsibility, maximising the good aspects and suggesting a solution to the problem.

2.4.1 Vocabulary task

Look through the Methodology sections in this unit and the Methodology or Experimental sections in your target articles. Underline or highlight all the words and phrases that you think could be used in the seven areas above.

A full list of useful language can be found on the next pages. This includes all the appropriate words and phrases you highlighted along with some other common ones. Read through them and check the meaning of any you don't know in the dictionary. This list will be useful for many years.

2.4.2 Vocabulary for the Methodology section

1. PROVIDE A GENERAL INTRODUCTION AND OVERVIEW OF THE MATERIALS/METHODS and GIVE THE SOURCE OF MATERIALS/EQUIPMENT USED

Some of the vocabulary you need for this is in the Introduction vocabulary list; for example, many of the verbs that describe what you did/used can be found there.

These verbs fall into three categories: the first includes general verbs related to academic research, such as *attempt, consider, conduct, determine, investigate, report, suggest, verify*, and most of these can be found in the Introduction vocabulary list. The second category contains verbs that specify what you did, such as *calculate, extract, isolate, formulate, incorporate, modify, plot, simulate,* and these can be found in the vocabulary list below. The third category includes verbs which are specific to your field and your research, but which are not useful in other fields, for example *clone, dissect, isotype, infuse*. Also try:

all (of)	(the) tests	is/are commercially available
both (of)	(the) samples	was/were acquired (from/by)
each (of)	(the) trials	was/were carried out
many (of)	(the) experiments	was/were chosen
most (of)	(the) equipment	was/were conducted
the majority(of)	(the) chemicals	was/were collected
	(the) models	was/were devised
	(the) instruments	was/were found in
	(the) materials	was/were generated (by)
		was/were modified
		was/were obtained (from/by)

		was/were performed (by/in) was/were provided (by) was/were purchased (from) was/were supplied (by) was/were used as supplied was/were investigated

Here are some examples of how these are used:

- **The impact tests used in this work** were a modified version of…
- **All reactions** were performed in a 27 ml glass reactor…
- **All cell lines** were generated as previously described in…
- **In the majority of the tests,** buffers with a pH of 8 were used in order to…
- **Both experiments** were performed in a greenhouse so that…
- The substrate **was obtained from** the Mushroom Research Centre…
- SSCE glass structures **were used** in this study to perform…
- The cylindrical lens **was obtained from** Newport USA and is shown in Fig. 3.
- **The material investigated** was a standard aluminium alloy; **all** melts were modified with sodium.
- Topographical examination **was carried out** using a 3-D stylus instrument.
- **The experiments were conducted** at a temperature of 0.5°C.

2. SUPPLY ESSENTIAL BACKGROUND INFORMATION

As well as describing standard procedures and techniques you may need to describe the equipment/apparatus or instrument you used or constructed. In order to do this accurately you need good control over the language of spatial location. Make sure you know how to use the words/phrases below. If you are not sure, write down the dictionary definition and use a concordance sampler (which you can find on the Internet) to see how they are used.

opposite	facing		
out of range (of)	within range (of)		
below	under	underneath	
above	over	on top (of)	
parallel (to/with)	perpendicular (to)	adjacent (to)	
on the right/left	to the right/left		
(to) bisect	(to) converge	(to) intersect	
near side/end	far side/end		
side	edge	tip	end
downstream (of)	upstream (of)		
boundary	margin	border	
on the front/back	at the front/back	in the front/back	in front (of)
higher/lower	upper/lower	inner/outer	
horizontal	vertical	lateral	
circular	rectangular	conical	
equidistant	equally spaced		
on either side	on both sides	on each side	
is placed	is situated	is located	occupies
is mounted (on)	is coupled (onto)	is fastened (to)	is positioned
is aligned (with)	is connected (to)	is fixed (to)	is embedded
extends	is surrounded (by)	is fitted (with)	is encased (in)
is attached to	is covered with/by	is joined (to)	

Here are some examples of how these are used:

- Porosity was measured **at the near end and at the far end** of the polished surface.
- The compression axis **is aligned with** the rolling direction…
- The source light was polarised **horizontally** and the sample beam can be scanned **laterally**.
- The mirrors **are positioned near** the focal plane.
- Electrodes comprised a 4 mm diam disk of substrate material **embedded in** a Teflon disk of 15 mm diam.

- The intercooler **was mounted on top of** the engine…
- The concentration of barium decreases **towards the edge**…
- Similar loads were applied to **the front and side** of the box…
- A laminar flow element **was located downstream of** the test section of the wind tunnel…

In which sentence(s) below was the table closest to the wall?

The table was placed	*against the wall.*
The table was placed	*next to the wall.*
The table was placed	*flush with the wall.*
The table was placed	*in contact with the wall.*
The table was placed	*right against the wall.*
The table was placed	*alongside the wall.*

In which sentence(s) below was the clock closest to the door?

The clock was located	*just above the door.*
The clock was located	*slightly above the door.*
The clock was located	*immediately above the door.*
The clock was located	*directly above the door.*
The clock was located	*right above the door.*

Note that *half as wide (as) = half the width (of); half as heavy (as) = half the weight (of); twice as long (as) = twice the length (of) and twice as strong (as) = twice the strength (of). Also note that with/having a weight of 20 kg = weighing 20 kg and with/having a width/length of 20 cm = 20 cm wide/long.*

3. PROVIDE SPECIFIC AND PRECISE DETAILS ABOUT MATERIALS AND METHODS

These verbs fall into three categories: the first includes general verbs used in academic research, such as *attempt, consider, conduct, determine, investigate, report, suggest, verify*, and these can be found in the Introduction vocabulary list (Section 1.4). The second category contains technical verbs which are specific to your field and your research, but which are not useful in other fields, for example *anneal, calibrate, centrifuge, dissect, fertilise, ionise, infuse*. These will not be given here because they are not generally useful. The third category is a set of less technical verbs that specify what was done or used, such as *calculate, extract, isolate, formulate, incorporate, modify, plot, simulate*. These usually occur in the passive *(was/were isolated)* and can be found in the vocabulary list below.

was adapted	was divided	was operated
was added	was eliminated	was optimised
was adopted	was employed	was plotted
was adjusted	was estimated	was positioned
was applied	was exposed	was prepared
was arranged	was extracted	was quantified
was assembled	was filtered	was recorded
was assumed	was formulated	was regulated
was attached	was generated	was removed
was calculated	was immersed	was repeated
was calibrated	was inhibited	was restricted
was carried out	was incorporated	was retained
was characterised	was included	was sampled
was collected	was inserted	was scored
was combined	was installed	was selected
was computed	was inverted	was separated
was consolidated	was isolated	was simulated
was constructed	was located	was stabilised
was controlled	was maintained	was substituted
was converted	was maximised	was tracked
was created	was measured	was transferred
was designed	was minimised	was treated
was derived	was modified	was varied
was discarded	was normalised	was utilised
was distributed	was obtained	

4. JUSTIFY CHOICES MADE

because* by doing…, we were able to chosen for (+ noun) chosen to (+ infinitive) for the purpose of (+ -ing or noun)** for the sake of (+ -ing or noun) in an attempt to (+ infinitive) in order to (+ infinitive) it was possible to (+ infinitive) offer a means of (+ -ing) one way to avoid... our aim was to (+ infinitive)	provide a way of (+ -ing) selected on the basis of… so as to (+ infinitive) so/such that so (+ -ing) thereby (+ -ing) therefore* thus (+ -ing) to (+ infinitive) to take advantage of which/this allows/allowed *etc.* with the intention of (+ -ing)

*See Section 1.2.2 for other examples of signalling language

**See box below for infinitives, -ing forms and noun forms of useful verbs. ∅ indicates that a noun form is not available or is not common in this type of structure

INFINITIVE	-ING FORM	NOUN FORM
achieve	achieving	achievement
allow	allowing	∅
assess	assessing	assessment
avoid	avoiding	avoidance
compensate for	compensating for	compensation for
confirm	confirming	confirmation
determine	determining	determination
enable	enabling	∅
enhance	enhancing	enhancement
ensure	ensuring	∅
establish	establishing	establishment
facilitate	facilitating	facilitation

guarantee	guaranteeing	guarantee
identify	identifying	identification
improve	improving	improvement
include	including	inclusion
increase	increasing	increase
limit	limiting	limitation
minimise	minimising	∅
obtain	obtaining	∅
overcome	overcoming	∅
permit	permitting	∅
prevent	preventing	prevention
provide	providing	provision
reduce	reducing	reduction
remove	removing	removal
validate	validating	validation

Here are some examples of how these are used:

- **To validate** the results from the metroscale model, samples were collected from all groups.
- The method of false nearest neighbours was selected **in order to determine** the embedding dimension.
- **For the sake of** simplicity, only a single value was analysed.
- **By partitioning** the array, all the multipaths could be identified.
- Zinc oxide was drawn into the laminate **with the intention of** enhancing delaminations and cracks.
- **The advantage of** using three-dimensional analysis was that the out-of-plane stress field could be obtained.
- **Because** FITC was used for both probes, enumeration was carried out using two different slides.
- The LVDTs were unrestrained, **so allowing** the sample to move freely.
- The cylinder was constructed from steel, **which avoided** problems of water absorption.

5. INDICATE THAT APPROPRIATE CARE WAS TAKEN

Most of the items in the box below are in adverb form, but they also occur in adjective form (*e.g. accurate*).

accurately	every/each	immediately	rigorously
always	exactly	independently	separately
appropriately	entirely	individually	smoothly
at least	firmly	never	successfully
both/all	frequently	only	suitably
carefully	freshly	precisely	tightly
completely	fully	randomly	thoroughly
constantly	gently	rapidly	uniformly
correctly	good	reliably	vigorously
directly	identical	repeatedly	well

Here are some examples of how these are used:

- A mechanical fixture was employed to hold the sonic horn **firmly** in place.
- After being removed, the mouse lungs were frozen and thawed **at least** three times.
- The specimen was monitored **constantly** for a period af 24 hours.
- They were then placed on ice for **immediate** FACS analysis.
- **Frequent** transducer readings were taken to update the stress conditions smoothly.
- The samples were **slowly and carefully** sheared to failure.

6. RELATE MATERIALS/METHODS TO OTHER STUDIES

There are three ways in which you might want to relate your materials/methods to those used in other studies.

Option 1: The procedure/material you used is **exactly the same as** the one you cite.

according to	as reported by/in	given by/in
as described by/in*	as reported previously	identical to
as explained by/in	as suggested by/in	in accordance with
as in	can be found in	the same as that of/in
as proposed by/in	details are given in	using the method of/in

**by* and *of* are usually followed by the name of the researcher or research team (*by Ross* or *using the method of Ross et al.*) and *in* is usually followed by the work (*in Ross et al. (2003)*). Another option is simply to give the research reference at the appropriate place in the sentence, either in brackets or using a superscript number.

Option 2: The procedure/material you used is **similar to** the one you cite.

a (modified) version of	(very) similar	(to) adapt
adapted from	almost the same	(to) adjust
based in part/partly on	essentially the same	(to) alter
based on	largely the same	(to) change
essentially identical	practically the same	(to) modify
in line with	virtually the same	(to) refine
in principle	with some adjustments	(to) revise
in essence	with some alterations	(to) vary
more or less identical	with some changes	
slightly modified	with some modifications	

Option 3: The procedure/material you used is **significantly different from** the one you cite.

a novel step was…	although in many ways similar	(to) adapt*
adapted from*	although in some ways similar	(to) adjust*
based on*	although in essence similar	(to) alter*
in line with		(to) change*

loosely based on partially based on partly based on*	with the following modifications/changes:	(to) refine* (to) revise (to) vary* (to) modify*

*as you can see, these can be used in **Option 2** as well as **Option 3.** When you use them in **Option 2** you may not need to state the differences between the procedure/material you used and the one you cite if they are not significant. In **Option 3** those differences or modifications are significant and you should say what they were, especially if they were modifications which improved the procedure/material.

Here are some examples of how these are used:

- Developmental evaluation was carried out using the Bayley Scales of Infant Development **(Bayley, 1969)**.
- The size of the Gaussians was adjusted **as in (Krissian *et al.*, 2000)**.
- The centrifuge is a **slightly modified** commercially available model, the Beckman J6-HC.
- The protein was overexpressed and purified **as reported previously.**[10,12]
- **A revised version of** the Structured Clinical Interview (4th edition)[6] was used.
- **We modified** the Du and Parker filter to address these shortcomings and we refer to this modified filter as the MaxCurve filter.
- In our implementation **we followed** Sato *et al.* (1998) by using a discrete kernel size.

7. INDICATE WHERE PROBLEMS OCCURRED

minimise problem	**minimise responsibility**	**maximise good aspects**
did not align precisely only approximate	limited by inevitably	acceptable fairly well

it is recognised that less than ideal not perfect not identical slightly problematic rather time-consuming minor deficit slightly disappointing negligible unimportant immaterial a preliminary attempt not significant	necessarily impractical as far as possible (it was) hard to (it was) difficult to unavoidable impossible not possible	quite good reasonably robust however* nevertheless* **talk about a solution** future work should… future work will…* currently in progress currently underway

*There is an interesting difference between the phrase *future work should* and the phrase *future work will*. When you write *future work should* you are suggesting a direction for future work and inviting the research community in your field to take up the challenge and produce the research. When you write *future work will* you are communicating your own plans and intentions to the research community and it should be understood that these plans and intentions belong to you — you're saying 'hands off!' to the rest of the research community and describing a research plan of your own

Here are some examples of how these are used:

- **Inevitably**, considerable computation was involved.
- Only a brief observation was feasible, **however**, given the number in the sample.
- **Although** centrifugation could not remove all the excess solid drug, the amount remaining was **negligible**.
- Solutions using (q = 1) differed **slightly** from the analytical solutions.
- Continuing research **will** examine a string of dc-dc converters to determine if the predicted efficiencies can be achieved in practice.
- **While** the anode layer was **slightly** thicker than 13 μm, this was a **minor deficit**.

2.5 Writing a Methodology Section

In the next task, you will bring together and use all the information in this unit. You will write a Methodology section according to the model, using the grammar and vocabulary you have learned, so make sure that you have both the model (Section 2.3.3) and the vocabulary (Section 2.4) in front of you.

In this unit you have seen the conventional model of the Methodology and the vocabulary conventionally used has been collected. Remember that when you write, your sentence patterns should also be conventional, so use the sentence patterns you have seen in the Methodology samples in this unit and in your target articles as models for your writing.

Follow the model exactly this time, and in future, use it to check the Methodology of your work so that you can be sure that the information is in an appropriate order and that you have done what your readers expect you to do in this section.

Although a model answer is provided in the Key, you should try to have your own answer checked by a native speaker of English if possible, to make sure that you are using the vocabulary correctly.

2.5.1 Write a Methodology section

The aim of this task is for you to learn how to describe what you did and used so that any reader can repeat exactly what you did and obtain exactly the same result as you obtained. Remember that you are expected to show that you carried out your work with due care and that you had good reasons for doing what you did. The message is: **This is exactly what I did, I did it carefully and I had good reasons for doing it in this way.**

To complete the task, imagine that you are writing up a research project which has carried out the first-ever attempt to cook chicken. Imagine that until now, everyone ate it raw. The task is to write a recipe for cooking chicken as if it were the Materials/Methods section of a research paper.

As an example, instead of starting by writing something like *Cut the chicken into four pieces*, you could perhaps start with an overview of the entire procedure, or by giving the source of your chicken. Did you obtain it from a supermarket? Was it supplied by a laboratory facility? You will need to say what you used to cut the chicken up; using an axe gives a very different result from using a 4 cm Sabatier steel knife! Instead of writing *Now put the chicken in a hot oven for about an hour and a half*, you should write something like: *The sample was then placed on a 300 × 600 mm stainless steel sheet and heated in a Panasonic E458 × 500 w standard fan-assisted oven for 90 minutes at 350°C.*

Don't worry if you don't know how to cook chicken — it doesn't matter if you report that you cooked it by boiling it in vodka, but you must give the exact quantity and the brand name of the vodka you used, so that your method and results can be replicated by someone else. Remember to use the passive voice and the appropriate tense.

The title of the research paper in which you report the new process is: **AN APPROACH TO THE PREPARATION OF CHICKEN.** The Introduction to your paper looks like this:

Introduction

Chicken preparation techniques are used in a range of applications both in homes and in restaurants. Chicken is easily available and can be locally produced in most areas; in addition it is easily digested and low in calories.[1]

Since Dundee's pioneering work reporting the 'natural' method of chicken preparation (Dundee et al., 1990) in which the chicken was killed and then eaten raw with salt, there have been significant innovations. Much work has been carried out in France in relation to improving the method of slaughtering chickens,[2] *whereas in the USA researchers have concentrated on improving the size of the bird.*[3,4] *The 'natural' method is widely used since the time required for the process is extremely short; however, some problems remain unsolved. The flavour of chicken prepared using the Dundee method is often considered unpleasant*[5] *and there is a well-documented risk of bacterial infection*[6] *resulting from the consumption of raw meat.*

The aim of this study was to develop a preparation method that would address these two problems. In this report, we describe the new method, which uses seasoning to improve the flavour while heating the chicken in order to kill bacteria prior to eating.

Now write the Methodology section of this paper. You should write approximately 250–400 words. If you get stuck and don't know what to write next, use the model and the vocabulary to help you move forward. Don't look at the Key until you have finished writing.

2.5.2 Key

Here is a sample answer. When you read it, think about which part of the model is represented in each sentence.

Two experiments were carried out using different combinations of seasoning and varying cooking temperatures. A 4.5 kg frozen organic chicken was purchased from Buyrite Supermarket. Buyrite only sell grade 'A' chickens approved by the Organic Farmers Association, thus ensuring both the homogeneity of the sample and the quality of the product. Seasonings were obtained from SeasonInc UK and were used as supplied.

According to the method described by Hanks *et al.* (1998), the chicken was first immersed in freshly boiled water cooled to a temperature of 20°C and was subsequently rinsed thoroughly in a salt solution so as to reduce the level of bacteria on the surface of the chicken.[7] In order to obtain two samples of equal size and weight for testing, the chicken was first skinned using a standard BS1709 Skin-o-matic; the flesh was then removed from the bone with a 4 cm steel Sabatier knife, after which it was cut into 3 cm-cubes, each weighing 100 g.

Two of the cubes thus obtained were randomly selected for testing. The cubes were dried individually in a Phillips R2D2 Dehydrator for 10 minutes. Immediately after removing each cube from the dehydrator it was coated with the selected seasoning mixture[8] and left to stand on a glass plate for 30 minutes at room temperature (16°C) in order to enhance absorption of the seasoning prior to heating. Seasoning quantities were measured used standard domestic kitchen scales and were therefore only approximate.

Each cube was then placed on an ovenproof dish and transferred to a pre-heated Panasonic Model 33KY standard electric fan-assisted oven at 150°C for 10 minutes. The product was removed from the oven and allowed to come to equilibrium, after which the cubes were assessed according to the TTS test developed by Dundee (Dundee, 1997).

Unit 3 ✏ Writing about Results

3.1 Structure

The title of this section varies in different disciplines, and also in different journals. Instead of Results, it is sometimes called 'Analysis' or 'Data Analysis'. The table below shows four options for the subtitles from this point until the end of the research paper.

Option 1	Option 2	Option 3	Option 4
Results *or* Data Analysis	Results *or* Data Analysis	Results and Discussion	Results *or* Data Analysis
Discussion	Discussion	Ø	Discussion and Conclusion(s)
Conclusion(s)	Ø	Conclusion(s)	Ø

In all cases this section reports your comments on what you found or observed, and if the subtitle contains the word Discussion (*i.e. Results and Discussion*), it includes some or all of the Discussion. As with the Methodology section, the best way to choose an appropriate subtitle is to look at the Guide for Authors of the journals you read regularly.

In most cases, the results of your work can be given in graphs, tables, equations or images. Why, then, should you bother to write a Results section? Why not simply provide good, clear graphs or tables with good, clear titles and perhaps a few notes underneath each? Thinking about these questions is a good way to begin to understand what you should be writing

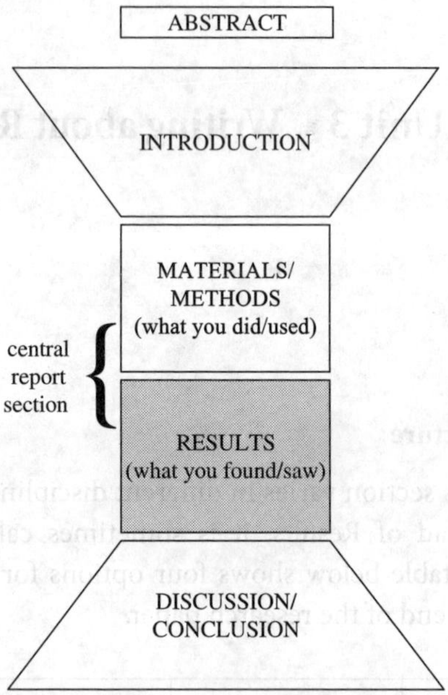

Fig. 1. The shape of a research article or thesis.

in this section. Almost everyone writes a Results section, so it is clear that some things cannot be achieved by just using tables, graphs or other images of your results. They can be achieved only by using words.

There are many reasons for writing a Results section. In the first place, some of your results may be more interesting or significant than others, and it is difficult to communicate this in a table or graph. Also, it is essential to relate your results to the aim(s) of the research. Thirdly, in some cases you may want to offer background information to explain why a particular result occurred, or to compare your results with those of other researchers. In addition, your results may be problematic; perhaps some experiments were not fully successful and you want to suggest possible reasons for this.

However, one of the most important reasons for writing a Results section rather than relying on graphs, tables and other images is that you must communicate your own understanding and interpretation of the results to your readers. Results do not speak for themselves; if they did, the tables or graphs of your results would be enough. Your readers do not have to agree with you but they need to know your opinion and understanding of your results.

So when we come to ask our three questions:

- How do I start the Results section? What type of sentence should I begin with?
- What type of information should be in this section and in what order?
- How do I end this section?

You already know that this section contains some comments on what you found or observed rather than just a description of your findings and observations, and this helps to answer the second question.

Read the Results section below. The title of the paper is: **A modelling approach to traffic management and CO exposure during peak hours**. Don't worry if the subject matter is not familiar to you or if you have difficulty understanding certain words, especially technical terms such as *median exposure*. Just try to get a general understanding at this stage and familiarise yourself with the type of language used.

Results

1 *Data obtained in previous studies*[1,2] *using a fixed on-site monitor indicated that travel by car resulted in lower CO exposure than travel on foot.* **2** *According to Figo et al. (1999), the median exposure of car passengers was 11% lower than for those walking.*[2] **3** *In our study, modelled emission rates were obtained using the Traffic Emission Model (TEM), a CO-exposure modelling framework developed by Ka.*[3] **4** *Modelled results were compared with actual roadside CO concentrations measured hourly at a fixed monitor.* **5** *Figure 1 shows the results obtained using TEM.*

6 *As can be seen, during morning peak-time journeys the CO concentrations for car passengers were significantly lower than for pedestrians, which is consistent with results obtained in previous studies.*[1,2] **7** *However, the modelled data were not consistent with these results for afternoon journeys.* **8** *Although the mean CO concentrations modelled by TEM for afternoon journeys on foot were in line with those of Figo et al., a striking difference was noted when each of the three peak hours was considered singly (Fig. 2).*

9 *It can be observed that during the first hour (H1) of the peak period, journeys on foot resulted in a considerably lower level of CO exposure.* **10** *Although levels for journeys on foot generally exceeded those modelled for car journeys during H2, during the last hour (H3) the levels for journeys on foot were again frequently far lower than for car journeys.*

11 *A quantitative analysis to determine modelling uncertainties was applied, based on the maximum deviation of the measured and calculated levels within the considered period.* **12** *Using this approach, the uncertainty of the model prediction for this study slightly exceeds the 50% acceptability limit defined by Jiang.*[7] **13** *Nevertheless, these results suggest that data obtained using TEM to simulate CO exposures may provide more sensitive information for assessing the impact of traffic management strategies than traditional on-site measurement.*

Before you begin to build a model, read the following section on grammar and writing skills.

3.2 Grammar and Writing Skills

This section deals with four language areas which are important in the Results section:

SEQUENCE
FREQUENCY
QUANTITY
CAUSALITY

3.2.1 Sequence

In order for other researchers to be able to repeat your work accurately and compare their results with yours, you need to be able to describe the order and time sequence of what you did and found in a very precise way. *Time sequence* means how long each step took and where it occurred in the sequence. You cannot use only *then* or *next*; these words tell your reader the order in which events occurred but they don't provide information about how long each event took, how soon the next event occurred or where it occurred in the sequence. A clear understanding of the time sequence will help your reader to picture it and repeat it for themselves.

The words and phrases that communicate sequence can be divided into eight groups.

1) The first group contains words or phrases which refer to events that occurred **before you began** your experiment/simulation or **before you began** observing your results:

> *It was apparent **beforehand** that a reduction in temperature would be a desirable outcome.*

2) The second group marks **the beginning** of the experiment/simulation or the **first** result you are describing:

> ***At the beginning** the temperature was stable, as predicted.*

3) The third group contains words/phrases which tell you the **order** in which events occurred but do not give any information about the time sequence:

> *The temperature increased to 49°C and **then** dropped to 30°C.*

In this case, the drop in temperature may have occurred quite soon after the temperature reached 49°C or it may have taken a long time; the word ***then*** only tells the reader the order in which these events occurred.

4) The fourth group is used to communicate that there was (only) **a short period of time** between two events:

> *The temperature increased to 49°C but **soon** dropped to 30°C.*

5) The fifth group communicates that the period of time between events was **long**, or that the event occurred **near the end** of the sequence:

> *The temperature increased to 49°C and **later** dropped to 30°C.*

6) The sixth group is extremely useful and important. It contains words and phrases that are used to communicate that events occurred **at the same time or almost at the same time**, or **during the same period,** and

therefore the items in this group are sometimes used to communicate a possible causal relationship between events:

*The temperature dropped sharply **when** we reduced the pressure.*

7) The seventh group marks **the end** of the sequence of events:

***At the end** there was a noticeable drop in temperature.*

8) The last group refers to events that occurred **after you finished** your experiment/simulation or **after you finished** observing the results:

*At the end there was a noticeable drop in temperature but it was decided **afterwards** to omit it from the input data.*

Here is a list of the words and phrases that communicate sequence:

after	firstly	previously
afterwards	formerly	prior to
as	immediately	secondly
as soon as	in advance	shortly after
at first	in the beginning	simultaneously
at that point	in the meantime	soon
at the beginning	in the end	straight away
at the end	initially	subsequently
at the same time	just then	then
at the start	lastly	to begin with
beforehand	later	to start with
before long	later on	towards the end
earlier	meanwhile	upon
eventually	next	when
finally	once	while
	originally	

Now put them into one (or more) of the appropriate groups. One example in each group has been entered in the box as a guide and some of the words or phrases can appear in more than one group.

1. *before the beginning*

beforehand

2. *the beginning or first step*

at the beginning

3. *steps/order*

then

4. *after a short while*

soon

5. *at a late/later stage; after a while/longer period*

later

6. *one point/period occurring almost or exactly at the same time as another*

when

7. *the end or last step*

at the end

8. *after the end*

afterwards

KEY

1. *before the beginning*

beforehand earlier formerly in advance	originally previously prior to

2. *at the beginning/first step*

at first at the beginning at the start firstly	in the beginning initially to begin with to start with

3. *steps/order*

after afterwards earlier next once	previously prior to secondly etc subsequently then

4. *after a short while*

before long shortly after	soon

5. *at a late/later stage; after a while/longer period*

eventually in time later	later on subsequently towards the end

6. *one point/period occurring almost or exactly at the same time as another*

as as soon as at that point at the same time immediately in the meantime just then	meanwhile simultaneously straight away upon + -ing when while

7. *at the end/last step*

at the end eventually	finally lastly

8. *after the end*

afterwards eventually	in the end later later on

3.2.2 Frequency

It is also important to communicate clearly how often a particular event or result occurred. If a particular result occurred *on every occasion* a test was carried out, then it is a very reliable result; if it *sometimes* occurred when the test was carried out, that is a less reliable result.

In the Methodology, if you write *x was done* without a frequency modifier, your reader may not be able to reproduce your method exactly. In the Results, if you write *x occurred* without a frequency modifier, your readers may not be able to compare their results with yours. Most importantly, readers may not be able to evaluate your results appropriately if they do not know how often a particular result occurred.

Frequency language has been arranged in the list below in order of frequency from 100% frequency (*on every occasion*) down to zero frequency (*never*). However, note that frequency language is often used in a subjective way: if something is said to occur *frequently*, this could be in relation to how often it was expected to occur. In other words, if previous research indicated that a particular result was unlikely to occur at all but in your study you find it on as many as 18% of occasions, you may consider that to be a frequent occurrence. On the other hand, if previous research indicated that something is very likely to occur but in your study you find it on only 57% of occasions, you may consider that to be relatively *rare*. Although frequency terms have an objective meaning, they can be used in a subjective way.

There is an identifiable mid-point in the list below: the phrase *as often as not* is used to express the fact that something occurred as often as it did not occur, *i.e.* with neutral frequency. Items appearing above that 50% mid-point express positive frequency and items appearing below it express negative frequency. However, apart from the first group and the last group, the items on the list cannot be quantified in terms of precise percentage frequency.

The list has been broken down into 11 groups, each of which contains items with more or less the same meaning.

1.	each/every time without exception on each/every occasion always invariably
2.	habitually as a rule generally normally usually
3.*	regularly repeatedly

4.*	frequently often commonly
5.	more often than not
6.	as often as not **(neutral frequency)**
7.	sometimes on some occasions at times
8.	occasionally now and then from time to time
9.	rarely seldom infrequently
10.	hardly ever barely ever almost never scarcely ever
11.	on no occasion not once at no time never

*The meanings of the items in Categories 3 and 4 are more flexible than those in the other categories.

Look at how the words/phrases in each group affect the meaning of a sentence. Imagine you want to find your supervisor on a Monday morning, and you want to know whether you should look for him in the library.

1. If he **always** goes to the library on Monday mornings you will find him there today.
2. If he **generally** goes to the library on Monday mornings you expect to find him there today and you will be surprised if he is not there.
3. If he **regularly** goes to the library on Monday mornings you will probably find him there today.
4. If he **often** goes to the library on Monday mornings there is a good chance that you may find him there today.
5. If he goes to the library **more often than not** on Mondays, you should start by looking for him there, but he may not be there today.
6. If he goes to the library **as often as not** on Monday mornings you may find him there today — or you may not. It's impossible to predict because the chances are equal; he goes there as often as he doesn't go there.
7. If he **sometimes** goes to the library on Monday mornings perhaps he will be there today (but you won't be surprised if he isn't there).
8. If he **occasionally** goes to the library on Monday mornings he might be there today but it's unlikely.
9. If he **rarely** goes to the library on Monday mornings he probably won't be there today (so don't bother to look for him there).
10. If he **hardly ever** goes to the library on Monday mornings he is not expected to be there today, and you would be surprised to find him there.
11. If he **never** goes to the Library on Monday mornings he won't be there today.

3.2.3 Quantity

Words have surprising power, and can encourage people to form strong impressions. Imagine, for example, that you are at a party and someone says to you: 'Come and meet my neighbour, he spent 10 years in prison for murder'. When you meet the neighbour his face may seem quite frightening. However, if you are told beforehand: 'Come and meet my neighbour, he gives a lot of money to poor people', he will probably look like a kind and caring person.

The language you use to describe your results has as much power as the tables and graphs themselves, perhaps even more. Look at the two curves in the figure below:

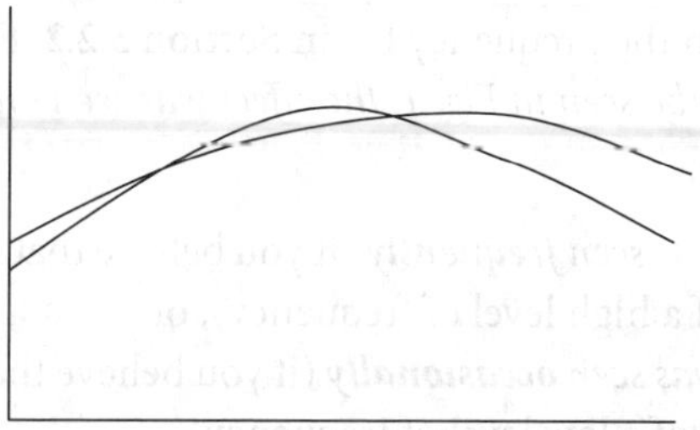

When you comment on this figure, if you write *As can be seen in the figure, the two curves are very similar,* the reader will focus on the similarity between the curves, and they will therefore seem similar. However, if you write *As can be seen in the figure, the two curves are noticeably different,* the reader will focus on the difference between them, and they will therefore seem different. The comments you make on your results tell your reader what you think about those results and influences the way readers perceive them.

Results do not speak for themselves! You can describe your results in numbers or percentages but those numbers or percentages are already visible to your reader in the graph or table; your reader needs to know what the numbers or quantities mean in order to understand them. For example, if the table or graph of your results shows that the effect you were looking for occurred in 23% of cases, you can communicate this as a strong result (*in **as many as** 23% of cases*) or a weak result (*in **only** 23% of cases*), but if you just write: *As can be seen in Fig. 1, the effect occurred in 23% of cases,* you have not added anything to what the reader can see for themselves.

Losing this opportunity to communicate what your results mean can cause problems. If you do not describe or comment on your results in words, the reader may perceive them differently from you. In other words, if you write *As can be seen in Fig. 1, the effect occurred in 23% of cases,* although you may have thought that was a high percentage, the reader may decide that *23% of cases* is low, or vice versa. This will have a damaging effect on the rest of your paper, in particular on your conclusions. You want your readers to accept your conclusions, and those conclusions should follow logically and naturally from your results. If you do not comment on your results so that the reader can share your understanding of them, s/he may see them differently. As a result, the conclusion you eventually derive from those results will not seem either natural or logical; in fact it can even seem surprising or rather strange to the reader.

One way to communicate your interpretation of the results is to use the language in the Frequency list in Section 3.2.2. For example, instead of writing *As can be seen in Fig. 1, the effect was seen on 23% of occasions*, you could write:

> *The effect was seen* ***frequently*** (if you believe that 23% of occasions is evidence of a high level of frequency), or
> *The effect was seen* ***occasionally*** (if you believe that 23% of occasions is evidence of a low level of frequency).

Another way to communicate your comments on the numbers, levels and quantities in the figure you are describing is to use **quantity language**. Quantity language can be used to replace numbers (*many*) or it can be used to comment on numbers (*as many as 45*). The words and phrases that communicate quantity can be divided into five groups.

1) The first group contains words or phrases which make the size/quantity look big:

> *A* ***considerable*** *amount of residue remained in the pipe.*

2) The second group contains words or phrases which make the size/quantity look small:

> ***Barely*** *23% of the residue remained in the pipe.*

3) The third group is used to **emphasise** how big/small/high/low the size/quantity is:

> *The amount that remained was* ***even higher/even lower*** *than predicted.*

4) The fourth group is used to communicate that the size/quantity is **similar/close** to another:

> ***Almost all/Almost half*** *of the residue remained in the pipe.*

5) The fifth group is useful when you need to say something about the quantity but you **do not want to commit yourself to an interpretation** of how big or small it was:

> ***Some*** *of the residue remained in the pipe.*

Here is a list of words/phrases which can be used in this way:

a great deal (of)
a few
a little
a number (of)
appreciable
appreciably (higher/lower)
approximately
as many as (*e.g.* 45)
as few as (*e.g.* 45)
at least
barely
below
by far
close (to)
considerable
considerably (higher/lower)
easily (over/under)
even (higher/lower)
exceptionally (high/low)
extremely (high/low)
fairly (high/low)
far (above/below)
few
fewer (than)
greater (than)
hardly
infinitesimal
in some cases
just
just (over/under)
less
little
marginal
marginally (higher/lower)
marked
markedly
moderate
more (than)
most
much
nearly
negligible
noticeable
noticeably
numerous
only
over (half/25%)
particularly
plenty
practically
quite
reasonably
relatively
significant
significantly
slight
small
so (high/low)
some
somewhat
substantial
substantially
to some extent
under
upwards of
virtually
well (under/over)

Now put them into the appropriate groups as described above. You can do this by imagining that the word/phrase is being used to describe the data in a figure, for example: *As can be seen in Fig. 1, a substantial amount of residue remained in the pipe.* Some of the words/phrases can be used in more than one category; for example, the word '*much*': *As can be seen in Fig. 1, much of the residue remained in the pipe* (Group 1) and *As can be seen in Fig. 1, the amount of residue remaining in the pipe was much lower than expected* (Group 3). One example in each group is given in the box as a guide.

1. words or phrases which **increase** the size/quantity

most

2. words or phrases which **reduce** the size/quantity

below

3. words or phrases which **emphasise** how big/small/high/low the size/quantity is

very

4. words or phrases which communicate that the size/quantity is **similar/close to another**

almost

5. words or phrases which communicate **a reluctance to commit oneself** to an interpretation of the size/quantity

some

KEY

1. words or phrases which **increase** the size/quantity:	
a great deal (of)	most
a number (of)	numerous
as many as (45)	over (half/25%)
appreciable	plenty
at least	much
considerable	substantial
greater (than)	significant
marked	upwards of
more (than)	
2. words or phrases which **reduce** the size/quantity:	
a few	little
a little	less
as few as 45	marginal
barely	negligible
below	only
few	slight
fewer (than)	small
hardly	under
infinitesimal	
3. words or phrases which **emphasise** how big/small/high/low the size/quantity is:	
appreciably	extremely (high/low)
by far	far (above/below)
considerably	particularly
easily (over/under)	so (high/low)
even (higher/lower)	substantially
exceptionally (high/low)	well (under/over)
4. words or phrases which communicate that the size/quantity is **similar/close to another**:	
approximately	little (*i.e.* close to none)
close (to)	nearly
few	practically

few (*i.e.* close to none)	slightly
just (over/under)	virtually
5. words or phrases which communicate **a reluctance to commit oneself** to an interpretation of the size/quantity:	
fairly	reasonably
in some cases	relatively
moderate	some
quite	somewhat
rather	to some extent

3.2.4 Causality

When you describe your results, you may want to indicate the relationships or connections between the events that you observed. Sometimes you may be able to state clearly that one event caused another, on other occasions you may want to say that one event caused another but you do not have proof of the causal connection between them. This section is designed to provide you with a variety of language options to represent your understanding of the relationships between the events you observed.

In some of the verbs or phrases, the position of the cause and the effect are fixed. For example, in *x produced y*, the subject, x, is the cause and the object, y, is the effect; in *x originated in y*, *x* is the effect and *y* is the cause. In others, however, such as *x is linked to y*, it depends on what the writer wishes to say; *x* could be either the cause or the effect, or the writer may simply want to indicate that *x* and *y* are connected in some way.

Some verbs/phrases in the list below communicate a clear/strong causal connection (*cause, produce, be due to*). Some refer to a partial cause (*be a factor in, contribute to*), some refer to the initial or first cause in a causal chain (*originate in, initiate*). There are also verbs and phrases in the list which communicate a weak causal connection (*be related to, link*). These are useful when you want to indicate that there is a connection between particular events or phenomena but perhaps are not certain which was the cause and which was the effect. You may not even be certain yourself that the connection between them is definitely a causal connection. These verbs and phrases are indicated by an asterisk (*) in the table. Not all these verbs can be used in the passive, but where the passive can be used, it is given in the list.

Note that:

- to be **a** cause of or **a** result of something implies that other factors were also involved, whereas to be **the** cause of or **the** result of something implies that it is the only cause or result.
- *x results from y* means *x is a consequence of y; whereas result in y* means *y is a consequence of x*

(be) a/the cause of	create/(be) created
(be) a/the consequence of	derive/(be) derived
(be) a factor in	effect/(be) effected
(be) a/the result of	elicit/(be) elicited
(be) due to	give rise to
accompany/(be) accompanied*	generate/(be) generated
account for/(be) accounted for	influence/(be) influenced
affect/(be) affected	initiate/(be) initiated
arise from	link/(be) linked*
ascribe to/(be) ascribed to	originate in
associate/(be) associated*	produce/(be) produced
attribute to/(be) attributed to	relate/(be) related*
bring about/(be) brought about	result from
cause/(be) caused	result in
come from	stem from
connect to/(be) connected to*	trigger/(be) triggered
contribute to	yield

Causal statements such as *x caused y* are risky because they may be disproved at a later stage. As a result, science writing has developed many ways of reducing the responsibility of the writer when making such statements. One of the ways which you have seen in the box above is to weaken the causal verb, so that instead of saying *x caused y* you may decide to say *x was linked to y*. Here are some other ways of reducing your risk and responsibility by 'softening' a causal statement:

You can begin with one of the following:

It appears that… It can/may* (therefore) be inferred/assumed that… It is (very/highly/extremely) probable/likely that… It is (widely/generally) accepted that… It is/may be reasonable to suppose/assume that … It is/may be thought/recognised/believed/felt that… It is/may/can be assumed that… It seems (very/highly) probable/likely that… It seems (likely) that… It would seem/appear that … The evidence points to the likelihood/probability that… The evidence suggests that… There is a clear/good/definite/strong possibility that… There is evidence to indicate that… This implies/seems to imply/may imply that… Apparently, (therefore), There seems to be/is a tendency to It is thought/said/recognised that	x caused y.

* There will be more about this type of verb, called *modal verbs*, in the next unit.

Another option is to add a frequency qualifier:

x often caused y
x commonly caused y
x rarely caused y

or a quantity qualifier:

x caused y in many cases
x caused y in some cases/to some extent
x caused y in virtually all cases

or a modal verb:

> *x may have caused y*
> *x might have caused y*
> *x could have caused y*

and you can use more than one in a sentence, depending on how 'weak' you want your claim to be — but be careful; if you add too many, the sentence may not mean very much at all:

> *The evidence points to the possibility that in many cases, x can contribute to certain types of y.*

3.3 Writing Task: Build a Model

3.3.1 Building a model

You are now ready to begin to build a model of the Results section by writing a short description of what the writer is doing in each sentence in the space provided below. The Key is on the next page. Once you have tried to produce your own model you can use the Key to help you write this section of a research article when you eventually do it on your own.

GUIDELINES

You should spend 30–45 minutes on this task. If you can't think of a good description of the first sentence, choose an easier one, for example Sentence 5, and start with that. Your model is only useful if it can be transferred to other Results sections, so don't include content words such as *traffic* or you won't be able to use your model to generate Results sections in your own field.

Remember that one way to find out what the writer is doing in a sentence, rather than what s/he is saying, is to imagine that your computer has accidentally deleted it. What changes for you, as a reader, when it disappears? If you press another key on the computer and the sentence comes back, how does that affect the way you respond to the information?

As mentioned in previous sections, another way to figure out what the writer is doing in a sentence — rather than what s/he is saying — is to look at the grammar and vocabulary clues. What is the tense of the main verb? What is that tense normally used for? Is it the same tense as in the previous sentence? If not, why has the writer changed the tense? What words has the writer chosen to use?

Don't expect to produce a perfect model. You will modify your model when you look at the Key, and perhaps again when you compare it to the way Results sections work in your target articles.

A modelling approach to traffic management and CO exposure during peak hours	**In this sentence, the writer:**
Results	
1 *Data obtained in previous studies[1,2] using a fixed on-site monitor indicated that travel by car resulted in lower CO exposure than travel on foot.*	1__________
2 *According to Figo et al. (1999), the median exposure of car passengers was 11% lower than for those walking.[2]*	2__________
3 *In our study, modelled emission rates were obtained using the Traffic Emission Model (TEM), a CO-exposure modelling framework developed by Ka.[3]*	3__________
4 *Modelled results were compared with actual roadside CO concentrations measured hourly at a fixed monitor.*	4__________
5 *Figure 1 shows the results obtained using TEM.*	5__________
6 *As can be seen, during morning peak-time journeys the CO concentrations for car passengers were significantly lower than for pedestrians, which is consistent with results obtained in previous studies.[2]*	6__________
7 *However, the modelled data were not consistent with these results for afternoon journeys.*	7__________

8 *Although the mean CO concentrations modelled by TEM for afternoon journeys on foot were in line with those of Figo et al., a striking difference was noted when each of the three peak hours was considered singly (Fig. 2).* **9** *It can be observed that during the first hour (H_1) of the peak period, journeys on foot resulted in a considerably lower level of CO exposure.* **10** *Although levels for journeys on foot generally exceeded those modelled for car journeys during H_2, during the last hour (H_3) the levels for journeys on foot were again frequently far lower than for car journeys.*	8__________ 9__________ 10__________
11 *A quantitative analysis to determine modelling uncertainties was applied, based on the maximum deviation of the measured and calculated levels within the considered period.* **12** *Using this approach, the average uncertainty of the model prediction for this study slightly exceeds the 50% acceptability limit defined by Jiang.[7]* **13** *Nevertheless, these results suggest that data obtained using TEM to simulate CO exposures may provide more sensitive information for assessing the impact of traffic management strategies than traditional on-site measurement.*	11__________ 12__________ 13__________

3.3.2 Key

In Sentences 1 and 2 '*Data obtained in previous studies[1,2] using a fixed on-site monitor indicated that travel by car resulted in lower CO exposure than travel on foot. According to Figo et al. (1999), the median exposure of car passengers was 11% lower than for those walking.[2]***' the writer refers to the findings and conclusions obtained by other researchers.**

Why not start by describing my results?

If you begin by describing individual results, the reader will need to build an overall scheme or pattern of your results by putting those individual results together. This is difficult for the reader to do; it is your job as a writer to arrange the information so that it is easy for a reader to process it. As with all subsections, therefore, it is more 'reader friendly' to start with some introductory material. When you start any new section or subsection in your work the first sentence(s) should provide a smooth transition for the reader between the new (sub)section and the previous one. There are two good ways to do this:

1. You can begin by offering an overview of the current section. This is a description of the overall pattern or trend of the results. If you start with individual results and your reader puts them together 'bottom-up' to create an overall picture of what happened, there is a risk that each reader may end up with a different picture of your results. It is preferable therefore to begin with general statements about what was found (*in most cases, generally speaking, overall*). Providing an overview enables you to show your reader the 'wall' before you begin to describe the 'bricks'. It is useful to remember that this type of general overview may need to be repeated when you move from one set of results to another.

2. You can begin by referring back to something from the previous section(s). For example, you can refer back to:

- the general aims of research in this area (which you mentioned in the Introduction)
- the specific research problem you are focusing on, or the aim of your project (which you mentioned in the Introduction)
- the methodology
- the original prediction or assumption to be tested (which you mentioned in the Introduction)
- the findings of other research in this area (which you mentioned in the Introduction)

Why should I re-state my specific research problem or the aim of my project here?

Eventually, in the Discussion, you will need to say to what extent your study (and in particular your findings) solves the problem or fulfils the aim(s) you set out in the Introduction. Your results should therefore be very closely related to your aim(s); in fact, when you examine your results carefully you may even decide to go back to the Introduction and redefine the original aim(s) in relation to the results you obtained.

Why should I re-state the original prediction or the findings of other research here?

Your results support, modify or contradict the original prediction, and they may support, modify or contradict the findings of other researchers. By repeating the original prediction or the findings of other researchers at the start of this section, your readers can see more clearly how your results relate to that prediction or those findings. Your readers will not remember the earlier parts of your paper as clearly as you do.

In Sentences 3 and 4 '*In our study, modelled emission rates were obtained using the Traffic Emission Model (TEM), a CO-exposure modelling framework developed by Ka.[3] Modelled results were compared with actual roadside CO concentrations measured hourly at a fixed monitor*.' **the writer refers back to his/her own methodology and adds more information about it.**

You may decide to refer to or summarise your methodology in your opening sentences. One reason for doing this is to highlight the important aspects of the materials, equipment or methodology you used to obtain your results. Another reason is to remind your readers of the methodology. You of course, remember it well — after all, it's your own research — but your readers don't share that familiarity. Also, extended details of the methodology are often given here rather than in the previous section, which may have included only the basic framework of the method. The more specific and complex the method, the more likely this is.

Background information is as common and as necessary here as elsewhere. In this case, information is provided about the instrument(s) or equipment used to obtain the results (*a CO-exposure modelling framework developed by Ka*[3]). Later on in the Results section you may need to provide more factual information in order to explain why a specific result occurred. For example, perhaps a result was obtained because of a particular property or characteristic of the materials used, in which case it would be appropriate for you to offer information about that property or characteristic to the reader. As always, it is better to offer slightly too much background factual information than too little. The wider the topic — and therefore the wider the readership — the more background information you should provide, so that all readers can understand why the results occurred as they did.

In Sentence 5 '*Figure 1 shows the results obtained using TEM.*' the writer invites the reader to look at a graph/figure/table *etc.*

Why do I need to invite the reader to look? Surely they will see the figure if they continue reading down the page.

What do you do when you are reading and you come to a sentence like this? You stop reading and take a look at the figure; you try to understand it or interpret the data you see in it; then you return to the text and keep that interpretation in mind when you carry on reading. If the data in the figure is very clear and has only one possible interpretation, it doesn't matter when you invite your reader to look at it. In this case, the results are very clear and easily interpreted, and so it is safe to let the reader view them before you comment on them. However, the data in many figures, tables and photographs can be interpreted in more than one way, in which case you should comment on the results in that figure before you invite the reader to take a look. If not, the reader may interpret it differently from you.

In Sentence 6 *'As can be seen, during morning peak-time journeys the CO concentrations for car passengers were significantly lower than for pedestrians, which is consistent with results obtained in previous studies.[2]'* **the writer refers to specific results and compares them with those obtained in another study, using subjective, evaluative language** (*consistent with*)**.**

Do I need to compare my results with those of other researchers?

One of the aims of this book is to make you aware of the difference between the kind of writing you produced before you began to do your own research and the kind of writing you want to produce now. Until now, you have probably written reports for people like your teachers or professors, who know more about your research topic than you. You have performed experiments or simulations that have already been performed by other researchers and the results were therefore predictable in most cases. Your only task was to describe the methods you used and the results you obtained to readers who already knew what methods you should use and what results you would obtain.

Now, however, things have changed, and in addition to reporting your results you should locate them on the 'research map' in your field. This means that you need to show your reader how and where your results fit in with the existing research picture, so you need to compare your results with those in the literature. You will develop this 'mapping' of your work more extensively in the Discussion, but in order to do it effectively, you need to first set your results against existing results.

What order should I present my results in?

The order in which you present your results to the reader is very important. You may be impatient to present your most important results, but it may be necessary to start by describing the results which underlie or lead to the more important ones.

Why do I have to use evaluative language — why not simply describe the results which are in the figure or table?

As stated earlier, results do not speak for themselves. You do not have to use evaluative language in every case; sometimes results can be given objectively, either numerically or in non-evaluative language. However, if you simply describe what is in the figure or table, you have not added anything to what the reader can see for themselves — so why bother? The comments you make on your results influence the way readers perceive them. As we noted earlier (in Section 3.2.3), if you write *As can be seen in Fig. 1, the two curves are very similar*, the reader will focus on the similarity between the curves; however, if you write *As can be seen in Fig. 1, the two curves are noticeably different*, the reader will notice the difference between them.

In Sentence 7 '*However, the modelled data were not consistent with parallel FOM measurements for afternoon journeys.*' **the writer offers a general statement about his/her results to begin a new paragraph.**

In this sentence, the writer is moving on to more interesting, controversial results, and communicates this to the reader by starting a new paragraph and using a signal (*However)* at the start of the sentence. As a writer, you always know which results are interesting or significant, but unless you communicate this to the reader by using a signal like this, all results will be perceived as having the same function and importance.

In Sentence 8 '*Although the mean CO concentrations modelled by TEM for afternoon journeys on foot were in line with those of Figo et al., a striking difference was noted when each of the three peak hours was considered individually (Fig. 2).*' **the writer refers to specific results and compares them to those obtained in another study, using language that comments on the result(s)** (*a striking difference*).

At some stage, you need to describe individual results in some detail, selecting results which are important, typical, or especially interesting.

Isn't a word like 'striking' considered too informal?

Definitely not. In science research writing, you do not normally use exclamation marks (!), even though you may feel that you want to if your results are very exciting. Instead, science writing uses a variety of words and phrases to achieve that 'wow!!' feeling, including *striking*. A full list of these can be found in the vocabulary section for the Discussion/Conclusion (Section 4.4).

Notice also that by starting the sentence with *Although*, the writer helps the reader to predict the function of the sentence correctly. The sentence could also have been written as follows:

> *The mean CO concentrations modelled by TEM for afternoon journeys on foot were in line with the FOM data* **but** *a striking difference was noted when each of the three peak hours was considered singly.*

but the reader would have had to wait until the middle of the sentence (*but*) to discover the function of the information in the first part of the sentence, and may have needed to 'loop back' through the first part of the sentence again in order to understand it. Signals are more useful when they occur early in the sentence.

> **In Sentences 9 and 10** '*It can be observed that during the first hour (H1) of the peak period, journeys on foot resulted in a considerably lower level of CO exposure. Although levels generally exceeded those modelled for car journeys during H2, during the last hour (H3) the levels for journeys on foot were again frequently far lower than for car journeys.*' **the writer selects specific results to describe in more detail, using language that comments on the results** (*considerably lower, generally, frequently far lower*).

Should I explain my results as well as present them?

That depends on the complexity of your results and the type of paper you are writing. Explanations can be given by providing background factual information to explain why a particular result occurred, for example, information about the properties of the material you are studying or the type of method you used. Make sure that you understand the difference between the *explanation* of a result (why it occurred as it did), the *evaluation* of a result (what the numbers mean) and the *implication* of a result (what the result suggests or implies). At this stage your explanations should be limited to fairly direct comments about your results; you will move on to broader explanations and implications in the Discussion/Conclusion.

How do I know which results to describe in detail? Why not describe all of them in detail?

If you describe all your results in equal detail they will seem to have the same level of importance. This is unlikely to be the case: some of your results are probably more significant than others, some are typical, and some are key results whereas others may be of more peripheral interest. However, your sentences are, in the end, simply black lines on a white page — the reader cannot hear your voice and so cannot hear you emphasising the importance of a particular result. You cannot print it in red and, as we have seen, you cannot even use an exclamation mark. So choosing to describe a specific result in detail communicates to your reader that you consider that particular result to be significant, worth highlighting or emphasising.

It is interesting to note that the best results are often described in such a way as to give the impression that they are typical results — look out for this in the papers you read. This is commonly done by stating a generalisation followed by *for example* and then the result: '*...the SFS results are in very good agreement with their FE counterparts; for example, at midspan the values are almost identical.*' Don't be ashamed of the need to persuade; if you proudly or shyly stick to simple descriptions of your results using 'naked numbers', your reader may be surprised by your conclusions because you have not said what those numbers mean. Your reader may not agree with you, but s/he needs to know what you think about your results.

In Sentence 11 '*A quantitative analysis to determine modelling uncertainties was applied, based on the maximum deviation of the measured and calculated levels within the considered period.*' **the writer refers to the method used to analyse the results.**

Why wasn't this included in the Methodology?

If you look at the Results sections in your target journals, you will be surprised by the amount of methodology included in this section.

The Methodology often only deals with the basic structure and components of the materials and methods. In such cases, most of the details are incorporated into the Results. This way of presenting information is becoming quite common in science journals.

> **In Sentence 12** '*Based on this approach, the average uncertainty of the model prediction for this study slightly exceeds the 50% acceptability limit defined by Jiang.*[7]' **the writer mentions a problem in the results and uses quantity language** *(slightly)* **to minimise its significance.**

Do I need to mention problems in the results? Won't it make the reader doubt my results?

As discussed in the Methodology, the opposite is true. Don't ignore problems in your results unless you are certain that the problems are insignificant and invisible. If your results are incomplete or some of them don't 'fit', you should mention this, minimise its importance if you can, and suggest possible reasons for the problem/offer a solution. Failing to mention a problem suggests that you aren't sufficiently expert to be aware of it, and this has a negative effect on your professional authority. By contrast, including a discussion of a problem in your work does exactly the opposite: it shows you to be fully in control of your research and able to evaluate it clearly. Furthermore, it provides you with an essential element for the Discussion/Conclusion: directions or suggestions for future research.

As with problems in the methodology, if you delay writing up until your results are all perfect, you may never get to publish it. So write it up as soon as your results are worth communicating; don't wait for perfection. Mention and acknowledge the problems or difficulties you encountered with your results while you are writing the Results section; it isn't appropriate to mention them for the first time when you are discussing suggestions for future work in the Discussion/ Conclusion.

So how can I talk about problems in the Results?

Use vocabulary that **minimises the problem**, **suggests possible reasons for it** and/or **offers a solution or a way forward**. In the example above, the writer acknowledged that there was a problem and minimised its effects (*slightly*). You can find examples of the language you will need to refer to imperfect or problematic results in the vocabulary list in Section 3.4.

> **In Sentence 13** '*Nevertheless, these results suggest that data obtained using TEM to simulate CO exposures may provide more sensitive information for assessing the impact of traffic management strategies than traditional on-site measurement.*' **the writer makes a reference to the implications and applications of the work s/he has done.**

Shouldn't that wait until the Discussion?

An examination of implications and applications is certainly one of the central areas of the Discussion, but most writers give some indication of what their results mean, *i.e.* the implications of their results, towards the end of the Results section. Once individual results have been described and discussed, the focus of the paper or thesis begins to open out and move away from the central 'reporting' section towards the conclusion. A sentence of this type is very common at this point, using verbs like *suggest* or *indicate*.

3.3.3 The model

Here are the sentence descriptions we have collected:

In Sentences 1 and 2	**the writer refers to the findings and conclusions obtained by other researchers.**
In Sentences 3 and 4	**the writer refers back to his/her own methodology and adds more information about it.**
In Sentence 5	**the writer invites the reader to look at a graph/figure/table *etc.***
In Sentence 6	**the writer refers to specific results and compares them with those obtained in another study, using subjective, evaluative language.**
In Sentence 7	**the writer offers a general statement about his/her results to begin a new paragraph.**
In Sentence 8	**the writer refers to specific results and compares them to those obtained in another study, using language that comments on the result(s).**

In Sentences 9 and 10	**the writer selects specific results to describe in more detail, using language that comments on the results.**
In Sentence 11	**the writer refers to the method used to analyse the results.**
In Sentence 12	**the writer mentions a problem in the results and uses quantity language to minimise its significance.**
In Sentence 13	**the writer makes a reference to the implications and applications of the work s/he has done.**

We can streamline these so that our model has FOUR basic components. Like the Methodology model, this is a 'menu' from which you select those items appropriate to your research topic and the journal you are submitting to.

1	REVISITING THE RESEARCH AIM/EXISTING RESEARCH REVISITING/EXPANDING METHODOLOGY GENERAL OVERVIEW OF RESULTS
2	INVITATION TO VIEW RESULTS SPECIFIC/KEY RESULTS IN DETAIL, WITH OR WITHOUT EXPLANATIONS COMPARISONS WITH RESULTS IN OTHER RESEARCH COMPARISON/S WITH MODEL PREDICTIONS
3	PROBLEMS WITH RESULTS
4	POSSIBLE IMPLICATIONS OF RESULTS

3.3.4 Testing the model

The next step is to look at the way this model works in a real Results section, and in that section (remember it may be called 'Analysis' or 'Data Analysis' instead) in the target articles you have selected. Here are some full-length Results sections from real research articles. Read them through, and mark the model components (1, 2, 3 or 4) wherever you think you see them. For example, if you think the first sentence corresponds to number 1 in the model, write 1 next to it *etc.*

Finite element modelling of sewer linings

4. NUMERICAL MODELLING

4.1. The finite element mesh

The cross-sectional geometries of the three egg-shaped linings are defined by the joining of two circles of differing diameters by slightly curved segments tangential to the circles. In the case of the St and Ch linings, the circles were osculating and with (275, 140 mm) and (330, 110 mm) radii, respectively; the Ce lining consisted of nonosculating circles of (250, 115 mm) radii, their nearest points from one another separated by a distance of 105 mm. Measurements were carried out on the cross-sections of the three lining types so as to determine the radii of curvature of the somewhat flat mid-section, a difficult task because of this flatness.

The thickness of the linings was found to vary along the cross-section; hence the mid-flat section tends to become slightly thicker than the rest of the cross-section. However, this is not accounted for in the model because it was observed that the thickness also varies along the length of the lining, and accurate measurements are not practically viable. So, a thickness of 6 mm is adopted for the St lining, 8 mm for the Ch, and a 10 mm thickness for the Ce lining. These thickness values are doubled at the hoop joints so as to simulate the actual junctions.

Due to symmetry about the vertical axis (*i.e.* the *y*-axis) of both loading and geometry, only half of the cross-section is analysed (see Fig. 2). Moreover, because of symmetry about the x-y plane, only half of the total lining length is considered (see Fig. 3). The cross-section of the lining which is under study is situated at the x-y plane of symmetry, located at a distance equal to half the total length of the lining. The reason for restricting the calculation of the stresses and deflections to this cross-section, is that full experimental data was only obtained at this lining location [8].

The element used in the analysis is an eight-noded isoparametric thin-shell element [6] with six degrees of freedom (*i.e.* three displacements and three rotations) at each node. Bending and membrane stresses are calculated at nodal points of individual elements and are then averaged for nodes which are common to more than one element.

Finally, the mesh adopted consists of 180 elements subdivided into nine elements in the longitudinal direction and 20 elements in the hoop direction (see Figs. 2 and 3). The subdivision of elements in the hoop direction consists of six elements at the invert of the lining, six elements at the mid-flat section, and eight elements at the top section of the lining (Fig. 2). The same mesh was used in the analysis of all three linings.

In addition to imposing the relevant displacement and rotation constraints along the two planes of symmetry that allow only one-quarter of the lining assembly to be analysed, and full fixity at the end of the pipe (*i.e.* all displacements and rotations are set to zero there), the boundary conditions corresponding to the five restraint set-ups were readily simulated by reference to the nodes situated at the joint section. Thus, whereas both displacements in the plane of the cross-section and the rotation about the longitudinal axis were suppressed at all mid-section nodes for BCI , no constraint for any of the degrees of freedom at the joint was imposed in the case of BC5 (see Fig. 1). For the three intermediate restraint set-ups, displacements of those nodes in contact with the wooden segment(s) were suppressed. All analyses were carried out by reference to a value of (uniform) suction pressure equal to 1 kN m^{-2}, and subsequent results, therefore, should be viewed with this value in mind.

4.2. Stanton and Staveley lining

As mentioned earlier, a thickness value of 6 mm was adopted in the analysis of the St pipe, this becoming equal to 12 mm for the elements located at the joint section. The mid-length of the pipe encompassed by the FE mesh was 1200 mm, as measured from the end of the lining assembly to the central cross-section monitored during testing and presently under study.

Figures 4 and 5 show a comparison between the experimental values of inner (hoop) strains and (transverse) deflections [8], and the ones stemming from the FE analysis for all five boundary cases.

(These and subsequent results are plotted against the vertical distance from the crown right up to the invert.) Here, the material properties used in the numerical model were selected from the lower range of values listed earlier, as this gave rise to a better correlation between the experiments and the FE analysis. These material properties for the lining were chosen as follows: E_h = 11.5 kN mm^{-2}, E_l = 5.5 kN mm^{-2}, ν_h = 0.29 and ν_l = 0.14, where, in order to be consistent with Betti's condition, $\nu_l \cdot E_h = \nu_h \cdot E_l$, it was decided to adjust the value of Poisson's ratio by taking the lower range of ν_l (ν_l being considered, in general, a more reliable test value than ν_h) at 0.14, and then working out ν_h at 0.29, a sensible approximation, as argued elsewhere [8].

It is noticed from the results of the analysis that the hoop strains and deflections resulting from the FE model follow similar patterns to the ones recorded from the vacuum rig tests. Moreover, it can be seen from Table 1 (which includes the maximum percentage error between the experimental results and the analytical predictions) that the FE model predicts reasonably well the critical values of deflection at the mid-flat section of the lining for the five test cases. This was also true for the critical values of inner strain (occurring at the invert of the lining) for test cases 3 and 5; however, the predicted values of critical bending strain were lower than their experimental counterparts for test cases 1, 2 and 4. This suggests that the FE model tends to predict reasonably well the behaviour of the lining in the absence of restraint set-ups, as can be shown from Table 1 (*i.e.* small percentage error in terms of critical strains and deflections for BC5). It will be shown later on that a more realistic simulation of the restraint set-ups can improve considerably the numerical predictions of the lining response.

4.3. Channeline lining

The half-length of the Ch pipe was 1170 mm and, again, its thickness (equal to 8 mm) was doubled for elements located at the joint section. Figures 6 and 7 show a comparison between the experimental results and the FE analysis using a lower range of values of isotropic material properties for reasons similar to those for the previous lining. These were chosen as follows: $E_h = E_l = 6.9$ kN mm^{-2} and $\nu_h = \nu_l = 0.33$.

Unlike the St pipeline, which was made up of one-piece linings, the Ch linings are two-piece segmental. As it is difficult to simulate realistically the behaviour of the longitudinal joint(s) in the case of segmental linings, the experimental results are still compared with the FE results using the mesh described earlier, even though such a mesh does not attempt to model the joint. From the outcome of this analysis, it can be shown that the response of the numerical model in terms of inner strains and deflections follows a similar pattern to that of the experimental results, as indicated in Figs. 6 and 7. At first sight, this would seem to imply that the longitudinal joint did in fact provide full shear and bending-moment continuity, so that the joint could be replaced by an equivalent continuous structure (*i.e.* stiffer than a hinge but well below fixity). This conclusion, however, need not follow since the shape of the bending-moment diagram is such that it changes sign in the vicinity of the longitudinal joint, so that, for this particular type of loading, the latter location is acted upon by relatively small bending action anyway. Therefore, on the basis of the bending strains, it is difficult to establish what the relative stiffness of the joint is, and it might seem reasonable to postulate a hinge (whether because of a flexible joint or simply due to the shape of the bending-moment diagram). On the other hand, it is shown from Table 1 that the numerical model predicts reasonably well the behaviour of the lining in terms of values of critical inner strains, but that it is not so good in terms of values of critical deflections. This fact would now suggest that the presence of the straight longitudinal joint might approach the effect of a hinge, thus allowing a larger rotation at the springings, with associated deflections at the flat mid-section of the lining, which are bigger than those obtained by the continuous-joint model in which the point of contraflexure does not occur exactly at the springing locations.

4.4. Celtite lining

The half-length of the pipe was 1200 mm and its thickness 10 mm (20 mm at the joint section).

Similarly to the previous cases, Figs. 8 and 9 show a comparison between the experimental observations and the FE-analysis results. Once again, a lower range of values of material properties is used; these properties are: $E_h = 13$ kN mm^{-2}, $E_l = 10$ kN mm^{-2}, $\nu_h = 0.10$ and $\nu_l = 0.13$. (It should be noted that, while ν_h was not obtained experimentally, it has been derived from Betti's condition.)

It may be seen that the patterns of inner strains and deflections from the numerical analysis are similar to their experimental counterparts. Since the Ce linings were segmental, remarks as for the Ch linings may be made regarding the straight longitudinal joint; namely, that, although the bending-moment diagram has a point of infection in the vicinity of the segmental joint, the larger percentage errors exhibited by both Channeline and Celtite linings, relative to the one-piece Stanton and Staveley lining, suggest that a line hinge may be more appropriate than the assumption of full continuity. Table 1 shows that good predictions in terms of the value of critical inner strain have resulted for boundary case 1 only, whereas in terms of critical deflection, such a conclusion applies for boundary cases 3 and 5, but not for the other three cases.

As for the previous two lining types, a stiffer response has resulted from the model. In addition to the presence of longitudinal joints, one must also point out that the deflection values recorded for the Ce lining during the experiments were small, and hence the error induced in the readings might have further affected their accuracy.

Observations of 2,4,6-trichlorophenol degradation by ozone

3. Results and discussion

3.1. Rate constants for the degradation of 2,4,6-TCP

In previous studies degradation rate constants have been established by undertaking ozonation experiments (Graham *et al.*, 2003) in the presence of a reference compound (Xiong and Graham, 1992a). The theoretical basis for this is as follows. The reaction of ozone with a solute M may be described by the following reaction scheme:

$$M + nO_3 \rightarrow \text{Moxid} \qquad (1)$$

where n is the stoichiometric factor for the number of ozone molecules consumed per molecule of M degraded. The stoichiometric factor for many organic substrates has been reported to vary in the range of 1–5 (Hoigne and Bader, 1983b), and values of 1 (Davis *et al.*, 1995) and 2 (Javier Benitez *et al.*, 2000a) have been proposed for 2,4,6-TCP. In practice it is usually assumed that the ozone reaction is first order with respect to ozone and solute M concentration, thus the rate law can be formulated as

$$-d[M]/dt = kM[O_3]\,[M] \qquad (2)$$

where kM is the rate constant for the degradation of solute M by O_3. Previous work by the authors (Chu and Wong, 2003) has confirmed that under conditions where the ozone concentration can be considered constant, the degradation of 2,4,6-TCP is first order with respect to its concentration. In this study, in order to determine the degradation rate constant kM, ozonation has been conducted with a mixture of a solute M1 (2,4,6-TCP) and a reference compound M2 having a known rate constant (*k*M2). According to Eq. (2), it can be shown that

$$-d[M_1]/dt = kM_1\ [O_3]\ [M_1] \quad (3)$$
$$-d[M_2]/dt = kM_2\ [O_3]\ [M_2] \quad (4)$$

Dividing Eq. (3) by (4), gives

$$\frac{d[M_1]}{d[M_2]} = \frac{kM_1\ [M_1]}{kM_2\ [M_2]} \quad (5)$$

Integration of Eq. (5) yields

$$\text{Ln}\ \frac{[M_1]_0}{[M_1]_1} = \frac{kM_1}{kM_2}\ \ \text{Ln}\ \frac{[M_2]_0}{[M_2]_t}. \quad (6)$$

Thus, a graph of $\text{Ln}\{[M_1]_0/[M_1]\}$ versus $\text{Ln}\{[M_2]_0/[M_2]\}$ yields a line whose gradient gives (kM_1/kM_2). Since the rate constant (kM_2) of M_2 is known, the value of kM_1 can be determined.

In these tests, the reference compound that was chosen was the herbicide atrazine (2-chloro-4-ethylamino-6-isopropylamino-1,3,5-triazine) since rate constants for this had been determined previously under the same conditions (Xiong and Graham, 1992a). Figure 1 shows the results of the ozonation tests in terms of the comparative degradation of 2,4,6-TCP and atrazine.

The calculated values for the rate constants for 2,4,6-TCP are shown in Table 1.

3.2. Reaction mechanism and dechlorination of 2,4,6-TCP

The rate constants shown in Table 1 indicate that the reactivity of 2,4,6-TCP is much greater at neutral pH than at low pH; this can also be seen in Fig. 2. This is partly explained by the much lower reactivity of undissociated 2,4,6-TCP with molecular ozone than in its substantially dissociated state at pH 7.5, and partly by the effect of hydroxyl radical-reactions at the higher pH. The latter effect is predominant at high pH and a previous study has shown a linear increase in pseudo first-order reaction rates with pH in the range of $7 < pH < 11$ (Chu and Wong, 2003). The results shown in Fig. 2 indicate that in the early stages of the reaction there is a large overall O_3:TCP reaction stoichiometry, thus, at a reaction time of 2 min, the stoichiometry is 89 and 47 mol O_3/mol TCP at pH 2 and 7.5, respectively.

An indication of the reaction mechanism during the ozonation of TCP is given by the extent of dechlorination. Recent studies (Han *et al.*, 1998; Chu and Wong, 2003) have suggested one specific mechanism whereby a hydroxyl group replaces one chlorine atom to form 2,6-dichloro-benzo-1,4-quinone (DCQ), as indicated in Fig. 3. From this it can be seen that the reaction leads to a reduction in solution pH through the formation of HCl. To investigate this, the ozonation of TCP at various initial pH levels was carried out without the use of a pH buffer and the results are summarized in Fig. 4. Evidence of significant proton generation was observed, and the rate of pH reduction increased with the initial pH of the solution. In the reaction scheme shown in Fig. 3, only one chlorine in the TCP is substituted by a hydroxyl group to produce DCQ, H^+ and Cl-. However, it is likely that further dechlorination of the remaining two chlorine atoms is possible under favourable conditions, such as at high pH where substantial hydroxyl radical generation occurs. It can be seen from Fig. 4 that for the ozone reaction at the initial pH of 8.17 the change in the solution pH suggested a proton generation equivalent to approximately 1 mM, which is stoichiometrically close to the total chlorine mass of the original TCP (0.88 mM). However, it should be noted that DCQ can be further degraded by cleavage of the aromatic ring leading to the formation of aliphatic products. These in turn may react with ozone to form organic acids, such as formic acid and acetic glycolic acids (Abe and Tanaka, 1997), thereby reducing the solution pH.

Direct measurements of chloride concentration were made during the buffered ozone tests and these are shown in Fig. 5. It can be seen that the degradation of TCP (each molecule having three chlorine atoms) generates significant chloride ions as one of the major products. The number of chloride ions released during the TCP degradation was found to range between 1.5 and 1.9 per degraded molecule of TCP, with the number in this range systematically increasing with the extent of TCP degraded. Since this is an average value for the reaction, it indicates that dechlorination is a major reaction mechanism and suggests that for a proportion of the TCP molecules there may be complete dechlorination.

Since it is speculated that at pH 7.5 a major part of the TCP reaction is via hydroxyl radical attack, it was thought that the generation of OH radicals may be limited in the presence of the carbonate buffer. To enhance the concentration of radicals the tests were repeated in the presence of hydrogen peroxide (H_2O_2/ HO_{-2} is a promoter of radical-type chain reactions) and the comparative results can be seen in Fig. 5. The concentration of hydrogen peroxide (15 mM) used in this case corresponded to a final H_2O_2:O_3 mass ratio of 0.8 g/g. It can be observed from Fig. 5 that there was only a very small enhancement (~5%) of TCP degradation when H_2O_2 was present during the ozonation. A similar effect was observed with a lower H_2O_2 concentration (7.5 mM) indicating that H_2O_2 concentration was not a sensitive factor. In contrast, there was a much greater production of chloride, with the number of chloride ions released during the TCP degradation ranging between 1.7 and 2.7 per degraded molecule of TCP, with the number in this range systematically increasing with the extent of TCP degraded. This considerably higher productivity of chloride ions, without a proportional increase in TCP degradation, suggests that the O_3/ H_2O_2 oxidising conditions are able to readily release chloride from intermediate compounds formed from the TCP degradation. It is assumed that the reaction with intermediate compounds is principally through OH· radicals, but direct H_2O_2 oxidation may also occur.

3.3. Degradation of TCP with humic acid

Humic substances (*e.g.* humic acid) are typically present in significant quantities (2–20 mg/l) in natural, and wastewaters. They have been shown to have a complex behaviour in ozone reactions in that they can act as initiators, promoters and scavengers of hydroxyl radicals, as well as being a substrate for molecular ozone reactions. Previous studies (*e.g.* with atrazine; Xiong and Graham, 1992b) have shown that relatively low concentrations of humic substances can substantially enhance the degradation of organic substrates, while higher concentrations can greatly reduce the degradation. A similar approach was used in this study in which the rate of degradation of 2,4,6-TCP was determined in the presence of different concentrations of HA. The results are shown in Fig. 6(a) and the TCP degradation curves were found to fit a pseudo-first-order model (R^2 = 0.997–0.999); the pseudo-first-order rate constants are shown in Table 2.

It can be seen in Table 2 that the peak degradation rate occurred in the presence of approximately 17 mg/l (as TOC) of HA, corresponding to a HA:TCP mass ratio of 0.43. The maximum increase in degradation rate is approximately 25% (cf. in the absence of HA). At a HA concentration of 56.1 mg/l, and presumably higher concentrations, the degradation rate decreased relative to that in the absence of HA. In comparison, Xiong and Graham (1992b) found that the optimal degradation of atrazine occurred at a mass ratio of humic substances (as DOC)-to-atrazine of 1.8. However, since only part of the humic substances would be HA, the corresponding HA:atrazine ratio would be lower, and therefore closer to the values shown in Table 2. A further comparison can be made between the enhanced TCP degradation caused by the presence of HA with that caused by hydrogen peroxide. Fig. 6(b) compares the TCP degradation rates for the optimal HA concentration (16.8 mg/l TOC) with 516 mg/l of H_2O_2 (≡ 15 mM; H_2O_2:O_3 final mass ratio of 0.8 g/g). If it can be assumed that the enhanced degradation rates in both cases is the result of increased OH· radical production, then it appears that the HA was more effective than the hydrogen peroxide.

An examination of the relationship between flowering times and temperature at the national scale using long-term phenological records from the UK

Results

Mean dates and standard deviations of dates, together with extreme early and late dates for all species, are shown in Table 1. It is apparent, and was noted by Jeffree (1960), that there is a bias towards extreme lateness for early-season species, which is less obvious, or even reversed, in later species. This is demonstrated in Fig. 2, where the 11 species for which 58 years of data are available are presented as box plots. The vertical dashed line represents the standardised mean of 0 days and asterisks represent extreme years. It is apparent that the species at the bottom of the figure (the earliest species) have more extreme late years and those at the top (the late species) have more extreme early years.

A summary of the stepwise regression models is given in Table 2. All but 1 of the 25 models was highly significant ($P < 0.001$). In general, coefficients for months close to the mean flowering date were negative, indicating that warmer temperatures promoted earlier flowering. At the same time, autumn coefficients were generally smaller and positive, indicating that some vernalisation requirement was necessary but also that the autumn influence was smaller than that of spring. Whilst the high number of comparisons suggests that some model terms would be included by chance alone, only the model for the autumn crocus looks peculiar, with a strong, positive influence of the previous autumn. The result of summing all of the regression coefficients together (see Table 2) suggests a response to warming of 2–10 days per °C, the greatest response being shown by the "midseason" species. Only the autumn crocus produces a positive response, suggesting that the remaining species would all flower earlier, sometimes substantially so, under climate warming.

Figure 3 shows the response of all 25 events to the single monthly CET to which they are most closely correlated. Although most regression models included multiple terms, the temperature for a single month is used for simplicity because display against several months simultaneously is not straightforward. Also in Fig. 3, the response of the autumn crocus to the June CET is shown, confirming that a negative response to summer temperature does exist, albeit apparently overwhelmed by the effect of the previous autumn. Finally, in Fig. 3, horse-chestnut flowering times are shown in relation to the mean March–May CET. A comparison with the simpler relationship with the April CET confirms that relationships are tighter when the temperatures of many months are considered together.

Now do the same in your target articles. We hope you obtain good confirmation of the model and can answer the three questions at the beginning of this section:

- How do I start this section? What type of sentence should I begin with?
- What type of information should be in this section, and in what order?
- How do I end this section?

3.4 Vocabulary

In order to complete the information you need to write this section of your paper you now need to find appropriate vocabulary for each part of the model. The vocabulary in this section is taken from over 600 research articles in different fields, all of which were written by native speakers and published in science journals. Only words/phrases which appear frequently have been included; this means that the vocabulary lists contain words and phrases which are considered normal and acceptable by both writers and editors.

In the next section we will look at vocabulary for the following seven areas of the model:

1. REVISITING THE RESEARCH AIM/EXISTING RESEARCH

This includes ways to remind the reader of what was said earlier. You should signal this (*As mentioned earlier,*) and then use the same words or phrases that you used originally — probably in the Introduction — to create an 'echo' for the reader.

2. GENERAL OVERVIEW OF RESULTS

This includes ways to introduce the general pattern or trend of your results so that the reader knows what to expect. Phrases like *in most cases* are common here.

3. INVITATION TO VIEW RESULTS

You can't always write *Figure 1 shows...* Figures and tables don't always *show* things; sometimes they *present* things or *summarise* things.

4. SPECIFIC/KEY RESULTS IN DETAIL

The language used to describe specific results includes both language which provides an *objective description* of the results *(lower)* and subjective, evaluative language (*significantly lower/slightly lower)*.

5. COMPARISONS WITH RESULTS IN OTHER RESEARCH

This includes the language you may need to compare your results with those of other researchers, to use their results to confirm/support yours and to compare your results with predictions, models or simulations. Phrases like *is in line with* and *correlate well with* are common here.

6. PROBLEMS WITH RESULTS

Remember that research is not necessarily invalidated by inappropriate results, provided they are presented in a conventional, professional way. Phrases such as *minor deficit* and *not within the scope of this study* will help you here.

7. POSSIBLE IMPLICATIONS OF RESULTS

Suggestions about what your results imply are a pivotal point in a paper, and signal the move towards the Discussion/Conclusion. Phrases such as *This indicated/suggested/implied that* and *It seems therefore that* are useful here; you can add some qualifying language as 'weakeners' if you want to reduce your risk and responsibility.

3.4.1 Vocabulary task

Look through the Results sections in this unit and in each of your target articles. Underline or highlight all the words and phrases that you think could be used in each of the seven areas given above.

A full list of useful language can be found on the following pages. This includes all the words and phrases you highlighted from the Results sections in this unit, together with some other common ones which you may have seen in your target articles. Underneath each list you will find examples of how they are used. Read through the list and check the meaning of any you don't know in the dictionary. This list will be useful for many years.

3.4.2 Vocabulary for the Results section

1. REVISITING THE RESEARCH AIM/EXISTING RESEARCH

as discussed previously, as mentioned earlier/before, as outlined in the introduction, as reported, in order to…, we examined… it is important to reiterate that… it is known from the literature that… it was predicted that… our aim/purpose/intention was to… since/because…, we investigated… the aforementioned theory/aim/prediction etc. to investigate…, we needed to… we reasoned/predicted that…

Here are some examples of how these are used:

- **Since** the angular alignment is critical, the effect of an error in orientation was **investigated experimentally.**
- **We reasoned that** an interaction in one network between proteins that are far apart in the other network may be a technology-specific artifact.
- **In earlier studies** attempts were made to establish degradation rate constants by undertaking ozonation experiments.
- **The main purpose of this work** was to test algorithm performance.
- **As mentioned previously, the aim of** the tests was to construct a continuous crack propagation history.
- **In this work, we sought to** establish a methodology for the synthesis of a benzoxazine skeleton.
- **It was suggested in the Introduction** that the effective stress paths may be used to define local bounding surfaces.

GENERAL OVERVIEW OF RESULTS

generally speaking,
in general,
in most/all cases,
in the main,
in this section, we compare/evaluate/present…
it is apparent that in all/most/the majority of cases,
it is evident from the results that…
on the whole
the overall response was…
the results are divided into two parts as follows:
using the method described above, we obtained…

Here are some examples of how these are used:

- **It is apparent that both** films exhibit typical mesoporous structures.
- **It is evident** that these results are in good agreement with their FE counterparts.
- **In general,** coefficients for months close to the mean flowering data were negative.
- Our confidence scores have an **overall** strong concordance with previous predictions
- **On the whole,** the strains and deflections recorded from the FE model follow similar patterns to those recorded from the vacuum rig tests.
- Levels of weight loss were similar **in all cases.**

INVITATION TO VIEW RESULTS

(data not shown)	Figure 1:	contains
(Fig. 1)		corresponds (to)
(see also Fig. 1)		demonstrates
(see Fig. 1)		displays
(see Figs. 1–3)		gives
according to Fig. 1		illustrates
as can be seen from/in* Fig.1		lists

as detailed in Fig.1 as evident from/in the figure as illustrated by Fig. 1 as indicated in. Fig.1 as listed in Fig.1 as shown in Fig.1 as we can see from/in Fig.1... can be found in Fig.1 can be identified from/in Fig.1 can be observed in Fig. 1 can be seen from/in Figure 1 comparing Figs. 1 and 4 shows that... data in Fig. 1 suggest that... displayed in Fig. 1 evidence for this is in Fig. 1 from Fig. 1 it can be seen that... inspection of Fig. 1 indicates... is/are given in Fig.1 is/are represented (*etc.*) in is/are visible in Fig. 1 in Fig. 1 we compare/present etc.... results are given in Fig.1 we observe from Fig. 1 that...	plots presents provides reports represents reveals shows summarises

**from* means 'can be deduced/concluded from' the figure/table whereas *in* means that it actually 'appears in' the figure/table

Here are some examples of how these are used:

- The stress data in Fig. 18 **indicate** a more reasonable relationship.
- Figure 3 **illustrates** the findings of the spatial time activity modelling.
- The overall volume changes are **reported** in Fig. 6(d).
- Similar results were found after loading GzmA into the cells **(data not shown)**.
- Typical cyclic voltammograms **can be seen in Fig. 1.**
- **Comparing Figs. 1 and 4** shows that volumetric strains developed after pore pressure had dissipated.

- The rate constants shown in Table 1 **demonstrate that** the reactivity is much greater at neutral pH.
- The results **are summarised** in Table 4.

SPECIFIC/KEY RESULTS IN DETAIL

When you look at your target articles, you will notice that it is harder to find examples of the language used to provide an *objective description* of the results than it is to find examples of the language used to provide a *subjective description* of the results, and that when it does occur, objective language is likely to be modified by a subjective 'add-on'. For example, a phrase like *slightly lower* or *much lower* is found more often than *lower* on its own. This is because, as mentioned earlier, an objective description of the results does not tell readers anything they don't already know from looking at the figure.

If you are having difficulty seeing the difference between *objective* and *subjective* language, remember that describing one level or quantity as being *higher* than another is an objective truth; to describe a level or quantity as *high* is a subjective evaluation.

(i) Objective descriptions

accelerate(d) all change(d) decline(d) decrease(d) delay(ed) drop(ped) exist(ed) expand(ed) fall/fell find/found increase(d)	is/are/was/were constant is/are/was/were different is/are/was/were equal is/are/was/were found is/are/was/were higher is/are/was/were highest is/are/was/were identical is/are/was/were lower is/are/was/were present is/are/was/were seen is/are/was/were unaffected is/are/was/were unchanged is/are/was/were uniform	match(ed) none occur(red) peak(ed) precede(d) produce(d) reduce(d) remain(ed) constant remained the same rise/rose sole/ly vary/varied

Numerical representations of percentages, levels, locations, amounts *etc.*, *i.e. a 2% increase* are, of course, also 'objective'.

Here are some examples of how these are used:

- There was a **lower** proportion of large particles present at lower pH.
- As can be seen in Fig. 8, there were **different** horizontal and vertical directional pseudofunctions.
- As can be seen, in the second trial the level of switching among uninformed travellers **was unchanged**.
- This kind of delamination **did not occur** anywhere else.
- The CTOA **dropped** from its initial high value to a constant angle of 4°.
- It eventually **levelled off** at a terminal velocity of 300 m/s.

(ii) Subjective descriptions

abundant(ly)	imperceptible(ibly)	remarkable(ably)
acceptable(ably)	important(ly)	resembling
adequate(ly)	in particular,	satisfactory
almost	in principle	scarce(ly)
appreciable(ably)	inadequate	serious(ly)
appropriate(ly)	interesting(ly),	severe(ly)
brief/(ly)	it appears that	sharp(ly)
clear(ly)	large(ly)	significant(ly)
comparable (ably)	likelihood	similar
considerable(ably)	low	simple(ply)
consistent(ly)	main(ly)	smooth(ly)
distinct(ly)	marked(ly)	somewhat
dominant(ly)	measurable(ably)	steep(ly)
dramatic(ally)	mild(ly)	striking(ly)
drastic(ally)	minimal(ly)	strong(ly)
equivalent	more or less	substantial(ly)
essential(ly)	most(ly)	sudden(ly)
excellent	negligible(ibly)	sufficient(ly)
excessive(ly)	noticeable(ably)	suitable(ably)
exceptional(ly)	obvious(ly)	surprising(ly)
extensive(ly)	only	tendency
extreme(ly)	overwhelming(ly)	the majority of
fair(ly)	poor(ly)	too + adjective
few		unexpected(ly)

general(ly) good high(ly) immense(ly)	powerful(ly) quick(ly) radical(ly) rapid(ly)	unusual(ly) valuable very virtual(ly)

PLUS all the rest of the language from the **frequency** and **quantity** lists (Sections 3.2.2 and 3.2.3).

*Here are some examples of how these are used (including examples from the **frequency** and **quantity** lists):*

- In **the majority of** cases, SEM analysis revealed a **considerably** higher percentage of fine material.
- As can be seen, the higher injection rate gave **satisfactory** results from all three methods.
- **Similar** behaviour was observed in all cases, with no **sudden** changes.
- It can be seen in Fig. 5 that the Kalman filter gives an **excellent** estimate of the heat released.
- The effect on the relative performance was **dramatic**.
- A **striking** illustration of this can be seen in Fig. 5.
- Comparing Figs. 4 and 5, it is obvious that a **significant** improvement was obtained in **the majority of** cases.
- It can be observed from Fig. 5 that the patterns are **essentially** the same in both cases.
- Figure 1 shows a **fairly** consistent material.
- It can be observed from Fig. 2 that there was **only** a **very small** enhancement when H_2O_2 was present.

COMPARISONS WITH OTHER RESULTS

If you are referring to other research, make sure that the location of the reference citation or number is accurate or other researchers may end up 'owning' your work. Remember that the right place for a research reference is not necessarily at the end of a sentence.

as anticipated	is/are better than
as expected,	is/are in good agreement
as predicted by…	is/are identical (to)
as reported by…	is/are not dissimilar (to)
compare well with	is/are parallel (to)
concur	is/are similar (to)
confirm	is/are unlike
consistent with	match
contrary to	prove
corroborate	refute
correlate	reinforce
disprove	support
inconsistent with	validate
in line with	verify

Many of these can be modified to match the level of certainty you want to express by adding expression such as:

It seems that
It appears that
It is likely that

(See Section 3.2.4 for more of these.)

Here are some examples of how these are used:

- **It is evident that** the SFS results obtained here are **in** exceptionally **good agreement with** existing FE results.
- Distributions are **almost identical** in both cases.
- Our concordance scores **strongly confirm** previous predictions.
- We see that the numerical model tends to give predictions that **are parallel to** the experimental data from corresponding tests.
- These results demonstrate that improved **correlation** with the experimental results was achieved using the new mesh.
- This is **consistent** with results obtained in [1].

- The results are qualitatively **similar** to those of earlier simulation studies.
- These trends are **in line with** the previously discussed structure of the of the ferrihydrite aggregates.

PROBLEMS WITH RESULTS

Remember that research is not necessarily invalidated by inappropriate results, provided they are presented in a conventional, professional way. The vocabulary below will help you to achieve this.

minimise the problem/focus on good results	**suggest reasons for the problem**
	may/could/might have been
(a) preliminary attempt	*or*
despite this,	was/were:
however,	
immaterial	beyond the scope of this study
incomplete	caused by
infinitesimal	difficult to (simulate)
insignificant	due to
less than ideal	hard to (control)
less than perfect	inevitable
(a) minor deficit/limitation	it should be noted that…
negligible	not attempted
nevertheless	not examined
not always reliable	not explored in this study
not always accurate	not investigated
not ideal	not the focus of this paper
not identical	not within the scope of this study
not completely clear	possible source(s) of error
not perfect	unavoidable
not precise	unexpected
not significant	unfortunately
of no consequence	unpredictable
of no/little significance	unworkable
only	unavailable
reasonable results were obtained	

room for improvement slightly (disappointing) (a) slight mismatch/limitation somewhat (problematic) (a) technicality unimportant	**offer a solution** further work is planned future work should... * future work will...* in future, care should be taken in future, it is advised that...

* Remember that the phrase *future work should* is used to suggest a direction for the research community, whereas *future work will* tells readers that this is your next project.

Here are some examples of how these are used:

- The correlation between the two methods was **somewhat** less in the case of a central concentrated point load.
- **It should, however, be noted that** in FE methods, the degree of mesh refinement may affect the results.
- **Nevertheless**, this effect is **only** local.
- Full experimental data was **only** obtained at one location.
- **Reasonable results were obtained** in the first case, and good results in the second.
- **It is difficult to** simulate the behaviour of the joints realistically.
- **Although** this was **not** obtained experimentally, it can be assumed to exist.
- **Future work should** therefore include numerical diffusion effects in the calculation of permeability.
- This type of control saturation is fairly common and therefore **of no significance**.

Here is an interesting table. It is supposed to be funny, but as you can see, it reflects a set of shared assumptions and a kind of 'code' used in the research community.

WHEN YOU WRITE THIS...	DO YOU MEAN THIS?
It has long been known that...	I can't remember the reference
This is of great theoretical and practical importance	This is interesting to me

It has not been possible to provide definite answers to these questions	The experiments didn't work out
High purity/very high purity/ extremely high purity	Composition unknown
Three of the samples were chosen for detailed study	The results of the others didn't make sense, so we ignored them
Typical results are shown	Only the best results are shown
Although some detail has been lost in reproduction, it appears to be clear from the original micro-graph that…	It is impossible to tell much from the original micrograph
Agreement with the predicted curve was: perfect excellent good reasonably good satisfactory fair not perfect as good as can be expected	Agreement with the predicted curve was: good fair poor very poor awful really awful imaginary non-existent
These results will be reported at a later date	I might get round to this sometime if I don't change careers
It is suggested that… It is believed that… It seems that…	I think that…

It is clear that much additional work is required before a complete understanding can be reached	I don't understand it
Unfortunately, a quantitative theory to account for these effects has not yet been formulated	Neither does anyone else
Correct within an order of magnitude	Wrong
It is hoped that this work will stimulate further research	This paper isn't very good, but neither is anyone else's
It is obvious	...but impossible to prove

POSSIBLE IMPLICATIONS OF RESULTS

At some stage (usually late) in the Results, it is appropriate to provide a general explanation or interpretation of what your results might mean. This is often the pivotal point in a paper, and signals the move towards the Discussion/Conclusion.

Choose your verb tense carefully. You can use the Present Simple or the Past Simple. Because the Present Simple is the tense used to express permanent truths and facts, using the Present Simple will give your sentence the status of a fact. Using the Present Simple therefore 'unlocks' your interpretation from your research and enhances its truth-value (*We found that x occurs, which indicate/suggests that y causes z*), If you are less confident, use the Past Simple (*We found that x occurred, which indicated/suggested that y caused z*).

Notice how many words from the list of vocabulary used to describe causal relationships are found here (see Section 3.2.4).

apparently	it is logical that
could* be due to	it is thought/believed that
could* be explained by	it seems that
could* account for	it seems plausible (*etc.*) that
could* be attributed to	likely
could* be interpreted as	may/might
could* be seen as	means that
evidently	perhaps
imply/implies that	possibly/possibility
indicate/indicating that	potentially
in some circumstances	presumably
is owing to	probably
is/are associated with	provide compelling evidence
is/are likely	seem to
is/are linked to	suggest(ing) that
is/are related to	support the idea that
it appears that	tend to
it could* be concluded that…	tendency
it could* be inferred that	unlikely
it could* be speculated that	there is evidence for
it could* be assumed that	we could* infer that
it is conceivable that	we have confidence that
it is evident that	would seem to suggest/indicate

**could* can be replaced by *may* or *might* or sometimes *can*; there is a grammar section on these modal verbs in the next unit.

Here are some examples of how these are used:

- **This suggests that** silicon is intrinsically involved in the precipitation mechanism.
- These curves **indicate that** the effective breadth is a minimum at the point of application of the load.
- Empirically, **it seems that** alignment is most sensitive to rotation in depth.
- Only the autumn crocus produced a positive response, **suggesting that** other species would flower earlier under climate warming.

- **It could be inferred** therefore that these **may have** reacted with ozone to form organic acids, such as formic acid.
- **This indicates that** no significant crystalline transformations occurred during sintering.
- **It is therefore speculated that** at pH 7.5 a major part of the reaction was via hydroxyl radical attack.
- **It is apparent that** this type of controller **may be** more sensitive to plant/model mismatch than was assumed in simulation studies.
- The results **seem to indicate that** this causes the behaviour to become extremely volatile.
- **It is evident that** the ψ at midspan increases with the increasing *r*.

In your native language you intuitively choose words and phrases which reflect exactly the appropriate strength of your claim and the level of risk you want to take in stating it. You need to be able to do this in English, both in this section and in the Discussion/Conclusion.

The sentence *We found that sunbathing causes cancer* expresses a very strong claim, but you can communicate a weaker form of it in many different ways. Here are some examples:

*We found that sunbathing **is related to the onset of** cancer.*
*We found that sunbathing **was related to** the onset of cancer.*
*We found that sunbathing **may have been related to** the onset of cancer.*
*We found **evidence to suggest that** sunbathing may have been related to the onset of cancer.*
*We found evidence to suggest that in **some cases/in many cases,** sunbathing may have been related to the onset of cancer.*
*We found evidence to suggest that in some cases, **excessive** sunbathing may have been related to the onset of **certain types of** cancer.*
***It is thought that** excessive sunbathing may **sometimes be considered as contributing to** the onset of certain types of cancer.*

3.5 Writing a Results Section

In the next task, you will bring together and use all the information in this unit. You will write a Results section according to the model, using the grammar and vocabulary you have learned, so make sure that you have the model (Section 3.3.3) and the vocabulary (Section 3.4) in front of you.

Throughout this unit you have seen that conventional science writing is easier to learn, easier to write and easier for others to read than direct translations from your own language or more creative writing strategies. You have learned the conventional model of a Results section and collected the vocabulary conventionally used. Your sentence patterns should also be conventional; use the sentences you have read in your target articles and in the Results printed here as models for the sentence patterns in your writing, and adapt them for the task.

Follow the model exactly this time. After you have practised it once or twice you can vary it to suit your needs. However, you should always use it to check Results sections you have written so that you can be sure that the information is in an appropriate order and that you have done what your readers expect you to do in this section.

Although a model answer is provided in the Key, you should try to have your own answer checked by a native speaker of English if possible, to make sure that you are using the vocabulary correctly.

3.5.1 Write a Results section

Imagine that you have just completed a research project which has been investigating a possible link between UFO (Unidentified Flying Object) sightings and earthquake prediction. The task in this exercise is to evaluate your data and findings as if you were writing the Results section of a research paper.

In your Introduction you stated that *various theories have been suggested for the increase in the number of UFO sightings immediately prior to an earthquake*. You claim that *it is possible that the increase in the number of sightings during the period immediately prior to an earthquake can be used to predict when an earthquake is likely to occur*.

In your Methodology, you described how you collected data and assessed it on the basis of certain criteria. Now you will present and evaluate this data in the Results.

Using Table 1 below, write the Results section of this paper. The title of your research paper is **The earthquake lights theory: an analysis of earthquake-related UFO sightings.** You should write approximately 300–400 words. If you get stuck and don't know what to write next, use the model and the vocabulary to help you move forward. Don't look at the model answer until you have finished writing. As usual, you can make up facts and references for this exercise.

Table 1: UFO sightings within 300 km of epicentre.

Country	UFO sightings for 7 days prior to earthquake	Earthquake magnitude	Average weekly UFO sightings	Description of UFO
Russia	55	3.2	11	Green ball of light
India	15	4.4	18	Fast-moving disc
Australia	120	6.0	30	White flashes of light
USA	275	5.6	75	Clusters of high-speed light
Canada	42	2.6	6	Blue-green egg-shaped object

3.5.2 Key

Here is a sample answer. When you read it, think about which part of the model is represented in each sentence.

Results

Based on the assumption that the timing of UFO sightings may be of significance,[2] the aim of this study was to investigate a possible link between the number of UFO sightings close to the epicentre during the period immediately prior to an earthquake, and the earthquakes that follow.

The process of evaluating UFO sightings is complex and time-consuming. Checks with police, air traffic control operators

Table 1: UFO sightings within 300 km of epicentre.

Country	UFO sightings for 7 days prior to the earthquake	Earthquake magnitude	Average weekly UFO sightings	Description of UFO
Russia	44	3.2	11	Green ball of light
India	15	4.4	18	Fast-moving disc
Australia	90	6.0	30	White flashes of light
USA	275	5.6	75	Clusters of high-speed light
Canada	48	3.6	6	Blue-green egg-shaped object

and meteorologists were performed. Where possible, witnesses were interviewed and videos of the area was examined in order to eliminate as many conventional explanations as possible, such as satellites, meteors, space debris and even bird flocks.[2,4,11] All the cases were documented using the procedure followed by Vader[4] and results are displayed in Table 1. The Richter scale[11] was used to measure magnitude.

It is evident from the results that overall, there was a marked increase in sightings during the seven days prior to the earthquake. These results are in line with those of Kenobi *et al.* (2004), who noted a mean fourfold increase worldwide.[9] In Russia and the USA, for example, the number of sightings increased approximately fourfold during the week preceding the earthquake, and in Canada the increase was even more dramatic. Although the number of sightings is low in Canada, this may have been due to a low national interest in UFOs; in addition, the earthquake took place in a sparsely-populated area of the country. It is significant that almost all the participants in each country gave exactly the same description of 'their' UFO, and that these descriptions were noticeably different from those obtained in other countries.

It appears from this evidence that the period immediately prior to earthquake activity was associated with an increase in the number of UFO sightings. However, this work represents only a preliminary attempt to establish such a link. The actual relationship between the two may be more complex; for example, it is possible that because a Star Wars film was released in the USA during the period under study, the number of sightings was higher that week without any real change in UFO activity. These results nevertheless suggest that monitoring UFO activity may provide useful input for earthquake prediction strategies.

Unit 4 Writing the Discussion/Conclusion

4.1 Structure

The title of this subsection varies from journal to journal. As noted in Unit 3, some journals end with a subsection titled *Discussion*, some end with a subsection titled *Results and Discussion* and others end with a subsection titled *Conclusions*. In the first two cases the elements which need to be included in the Discussion are similar. Where there is a *Conclusions* section, it is short, usually comprising one or two paragraphs focusing on specific aspects of the Discussion.

The graphic representation at the beginning of each unit is symmetrical because many of the elements of the Introduction occur again in the Discussion/Conclusion in (approximately) reverse order. The Introduction moves from a general, broad focus to the narrower 'report' section of the paper, and the Discussion/Conclusion moves away from that narrow section to a wider, more general focus. The Discussion looks back at the points made in the Introduction on the basis of the information in the central report section.

Let us look again at the four components of the Introduction:

1	ESTABLISH THE IMPORTANCE OF YOUR FIELD PROVIDE BACKGROUND FACTS/INFORMATION (possibly from research) DEFINE THE TERMINOLOGY IN THE TITLE/KEY WORDS PRESENT THE PROBLEM AREA/CURRENT RESEARCH FOCUS

2	PREVIOUS AND/OR CURRENT RESEARCH AND CONTRIBUTIONS
3	LOCATE A GAP IN THE RESEARCH DESCRIBE THE PROBLEM YOU WILL ADDRESS PRESENT A PREDICTION TO BE TESTED
4	DESCRIBE THE PRESENT PAPER

When you started the Introduction, you helped your readers move *into* the research article by establishing that the topic was a significant topic, providing background information and so on. Following the same pattern in reverse, you end the Discussion/Conclusion by helping your readers move *out* of the article.

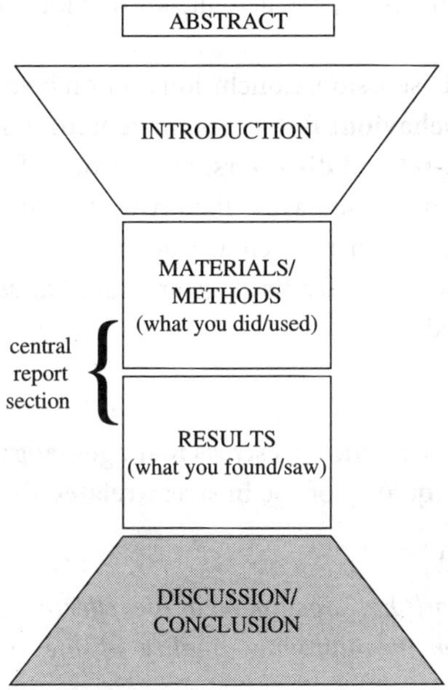

Fig. 1. The shape of a research article or thesis.

In the Introduction, you wrote about the work of other researchers, creating a kind of research map for your readers so that they could see what type of work existed in this field; in the Discussion/Conclusion you locate your study in relation to that research map. You then went on in the Introduction to locate a gap in the research or describe a problem associated with existing research; in the Discussion/Conclusion, you are expected to say to what extent you have responded to that gap or solved that problem. At the end of the Introduction you wrote about *the present paper*, creating an interface with the content of your own work so that you could move the reader on to the central report section of your paper; in the Discussion/Conclusion, as we will see, it is common to begin by revisiting some aspect of your work, so as to create that interface in reverse and enable you to move away from the central report section.

So as you can see, when we come to ask our three questions:

- How do I start the Discussion/Conclusion section? What type of sentence should I begin with?
- What type of information should be in this section, and in what order?
- How do I end this section?

although you may think that you have no idea of how to write the Discussion/Conclusion, you actually know a lot about what to include and in what order.

Read the Discussion/Conclusion section below. The title of the paper is: **Cognitive-behavioural stress management (CBSM) skills and quality of life in stress-related disorders.** Don't worry if the subject matter is not familiar to you or if you have difficulty understanding some of the words, especially technical terms such as *Cognitive-Behavioural*. Just try to get a general understanding at this stage and familiarise yourself with the type of language used.

Cognitive-behavioural stress management (CBSM) skills and quality of life in stress-related disorders.

Discussion

1 *Prior work has documented the effectiveness of psychosocial intervention in improving quality of life (QoL) and reducing* *stress in patients suffering from various disorders; Epstein,*[18] *for example, reports that orthopedic patients participating in a two-week multimedia intervention programme improved across several QoL indices, including interpersonal conflict and mental health.* **2** *However, these studies have either been short-term studies or have not focused on patients whose disorder was stress-related.* **3** *In this study we tested the extent to which an extended three-month stress management programme improved QoL among a group of patients being treated for stress-related skin disorders such as eczema.*

4 *We found that in virtually all cases, participation in our three-month stress management programme was associated with substantial increases in the skills needed to improve QoL.* **5** *These findings extend those of Kaliom, confirming that a longer, more intensive period of stress-management training tends to produce more effective skills than when those skills are input over a shorter period via information transfer media such as leaflets and presentations (Kaliom et al., 2003).* **6** *In addition, the improvements noted in our study were unrelated to age, gender or ethnic background.* **7** *This study therefore indicates that the benefits gained from stress-management intervention may address QoL needs across a wide range of patients.* **8** *Most notably, this is the first study to our knowledge to investigate the effectiveness of extended psychosocial intervention in patients whose disorder is itself thought to be stress-related.* **9** *Our results provide compelling evidence for long-term involvement with such patients and suggest that this approach appears to be effective in counteracting stress that may exacerbate the disorder.* **10** *However, some limitations are worth noting.* **11** *Although our hypotheses were supported statistically, the sample was not reassessed once the programme was over.* **12** *Future work should therefore include follow-up work designed to evaluate whether the skills are retained in the long term and also whether they continue to be used to improve QoL.*

4.2 Grammar and Writing Skills

This section deals with a complex language area which is important in the Discussion section: MODAL VERBS.

The modal verbs that are commonly used in science writing are *may, might, could, can, should, ought to, need to, have to* and *must*. Modal verbs which are not used in formal science writing, such as the use of *can* or *may* for 'permission' (*e.g. Can I borrow your pen?)* are not discussed here.

Modal verbs are often used to modify the 'truth value' of a sentence. In a sentence like:

*The drop in pressure **was** due to a crack in the pipe.*

there is no modal verb — you are telling your reader what caused the drop in pressure, and you have empirical evidence to prove it. If, however, you write that:

*The drop in pressure **may have been** due to a crack in the pipe.*

you are offering a possible cause for the drop in pressure; perhaps it was due to a crack in the pipe — and perhaps not. If you write

*The drop in pressure **must have been** due to a crack in the pipe.*

you are saying that you are virtually certain that the drop in pressure was caused by a crack in the pipe, but you do not actually have evidence to prove it. Being certain that something is true and knowing it to be true are not the same thing at all. For example, you would not look at your watch and say 'It must be ten o' clock' or 'I'm certain it is ten o' clock' — you would simply say 'It is ten o' clock'. You would be more likely to say 'It must be ten o' clock' when you aren't wearing a watch — in other words, if you aren't really sure, or you lack empirical evidence. Although using the modal verb *must* seems to give the sentence more power, it also communicates an absence of proof.

Modal verbs are particularly useful in the Results and Discussion sections. In these sections you are writing about the reasons, interpretations and implications of your results and you often need to communicate that something is a **possible** reason, or an **obvious** interpretation or a **probable** implication. Here is a typical sentence from a (combined) Results and Discussion section:

> *The kinetics* **can** *be described by these equations, suggesting that the electrons are transferred directly. This* **might** *involve a supercharge mechanism, but the data* **could** *also be described by electron transfer via a hopping mechanism.*

As a short exercise to start thinking about the way these verbs work, try to match the modal verbs in Column A to their meanings in Column B. Most of the modal verbs can be used for more than one meaning.

A	B
	ABLE/CAPABLE He … go home by himself. *(He is able to go home by himself.)*
1. SHOULD 2. MUST	POSSIBLE/OPTIONAL He … go home. *(It is possible that he will go home.)*
3. CAN 4. OUGHT TO	PROBABLE/LIKELY He … be home soon. *(He will probably be home soon.)*
5. MAY 6. COULD	VIRTUALLY CERTAIN He … be at home. *(It is virtually certain that he is at home.)*

A	B
7. NEED TO 8. MIGHT	ADVISABLE He… go home. *(I advise him to go home.)*
9. HAVE TO	NECESSARY He… go home. *(It is necessary for him to go home.)*

Now check your answers with this Key:

CAN	ABLE/CAPABLE *(He can go home by himself.)*
MAY MIGHT COULD CAN	POSSIBLE/OPTIONAL *(He may/might/could/can be at home,)*
SHOULD OUGHT TO	PROBABLE/LIKELY *(He should/ought to be home soon.)*
MUST HAVE TO	VIRTUALLY CERTAIN *(He must/has to be at home.)*
SHOULD OUGHT TO	ADVISABLE *(He should/ought to go home.)*
MUST NEED TO HAVE TO	NECESSARY *(He must/needs to/has to go home.)*

There are two reasons why these verbs are difficult to use. First of all, as you can see, some modal verbs have more than one meaning. Therefore when you use a verb like *should*, make sure you know whether you mean that something is *likely* to happen (or to have happened), or whether you mean that it is *advisable* for it to happen.

Second, most modal verbs do not follow standard grammar rules. Some disappear and others change their meaning in the negative, or in a different tense. For example, *He must go home* means the same as *He has to go home*, but *He must not go home* means that he is not allowed to go home, which is not the same as *He doesn't have to go home.*

Here is a table showing how each of these modal verbs works in the past tense and in the negative, with examples. The table does not deal with every possible use of every modal verb. The modal verb *may*, for example, is also used to request permission (*May I borrow your pen?*) but you are unlikely to need this in science research writing. As the function of this book is to help you write an accurate and acceptable research article, the information in this section has been limited to what you need.

1. ABILITY/CAPABILITY

Present Simple	**CAN**	This software **can** distinguish between different viruses.
Present Simple negative	**CANNOT**	Until 18 months a child **cannot** use symbols to represent objects.
Past Simple	**COULD** **COULD HAVE**	It was found that the gun **could** shoot accurately even at 300 meters. If we had extended the time period we **could have** produced more crystals.
Past Simple negative	**COULD NOT** **COULD NOT HAVE**	1n 1990, 80% of households **could not** receive digital television. The subjects reported that they **could not have** fallen asleep without medication.

Notes:

- The modal verb **can** only forms these two tenses when it refers to ABILITY or CAPABILITY. If you need other tenses, you will need to switch to **be capable of** or **be able to,** *i.e. It is believed that this software* ***will*** *eventually* ***be capable of*** *distinguishing between different viruses.*
- **could** means 'was generally capable of doing/able to do something in the past', whereas **was able to** is used in relation to specific past events or past occasions, *i.e. The result suggests that in this case, the viruses* ***were able to*** *multiply freely.* If you're not sure whether to use **can** or **be able to,** use **be able to** — it's safer.

2. POSSIBILITY/OPTIONS

Present Simple	**MAY** **MIGHT** **COULD** **CAN**	A rubber seal **may/might/could/can** be useful at this location.
Present Simple negative	**MAY NOT** **MIGHT NOT** (but not **COULD NOT** or **CANNOT**)	A rubber seal **may not/might not** be useful at this location.
Past Simple	**MAY HAVE** **MIGHT HAVE** **COULD HAVE** (but not **CAN HAVE**)	The fall in pressure **may have been/might have been/could have been** caused by leakage.
Past Simple Negative	**MAY NOT HAVE** **MIGHT NOT HAVE** (but not **COULD NOT HAVE** or **CANNOT HAVE**)	The fall in pressure **may not have been/might not have been** caused by leakage.

Notes:

- The word 'well' is sometimes added to communicate a stronger belief in the possibility: *This* ***may well*** *be due to leakage.*

- **might** is slightly weaker than **may**.
- Interestingly, **can not** and **cannot** don't mean the same thing at all! **can not** means *possibly not* in the same way as **may not** or **might not**, but it is rarely used except in structures such as 'this **can not** only damage the sample, it **may** even destroy it completely'. **cannot**, on the other hand, means something completely different: it doesn't mean *possibly not*, it means *impossible*. **could not**, **cannot**, **could not have** and **cannot have** all fall into this category. In sentences like:

  *We realise that this **cannot** be due to a change in pressure.*
  *We realised that this **could not** be due to a change in pressure.*
  *We realise that this **cannot have been** due to a change in pressure.*
  *We realised that this **could not have been** due to a change in pressure.*

  The writer is not saying 'possibly not', s/he is saying 'impossible'.

3. PROBABILITY/BELIEF/EXPECTATION

Present Simple	**SHOULD** **OUGHT TO**	The material **should** remain stable if it is kept below 30°C.
Present Simple negative	**SHOULD NOT** **OUGHT NOT TO**	The material **should not** decompose unless heated above 30°C.
Past Simple	**SHOULD HAVE** **OUGHT TO HAVE**	By the time the cobalt is added, the crystals **should have** dissolved.
Past Simple negative	**SHOULD NOT HAVE** **OUGHT NOT TO HAVE**	This was unexpected; the material **should not have** decomposed at this temperature.

Note: Although **ought to** means the same as **should**, it is less common in science writing, so examples have not been given.

4. VIRTUAL CERTAINTY

Present Simple	**MUST** **HAVE TO**	Our results indicate that contamination **must** be due to the presence of sea water in the pipe.
Present Simple negative	**CANNOT**	It is clear that contamination **cannot/could not** be due to the presence of sea water in the pipe.
Past Simple	**MUST HAVE**	Our results indicate that contamination **must have been** due to the presence of sea water in the pipe.
Past Simple Negative	**CANNOT HAVE** **COULD NOT** **COULD NOT HAVE**	It was clear that contamination **could not be/ cannot have been/could not have been** due to the presence of sea water in the pipe.

Notes:

- 'virtual certainty' modals communicate the fact that no other explanation is possible.
- **have to** is less common in science writing, so examples have not been given.
- **must not** means 'not allowed/permitted', it doesn't mean 'not possible'.

To separate Categories 2, 3 and 4, imagine that it normally takes Professor Windblast about 20 minutes to walk home from his laboratory. Has he arrived home yet? Well, you won't know unless you call his house and speak to him, but

- if he left the lab 18 minutes ago, he **may/might/could** be home by now *(possibly)*

- if he left 30 minutes ago, he **should/ought to** be home by now (*probably*)
- if he left 50 minutes ago, he **must** be home by now (*almost certainly*)
- if he left 5 minutes ago he **cannot** be home yet (*almost certainly not*)

5. ADVICE/ OPINION

Present Simple	**SHOULD** **OUGHT TO**	The apparatus **should** be disconnected from the mains during repairs.
Present Simple negative	**SHOULD NOT** **OUGHT NOT TO**	This material **should not** be exposed to sunlight
Past Simple	**SHOULD HAVE** **OUGHT TO HAVE**	The apparatus **should have been** disconnected from the mains during repairs.
Past Simple Negative	**SHOULD NOT HAVE** **OUGHT NOT TO HAVE**	This material **should not have been** exposed to sunlight

Notes:

- Although **ought to** means the same as **should**, it is less common in science writing, and that is why examples have not been given.
- **should have /ought to have** usually refer to something that didn't occur and **should not have/ought not to have** usually refer to something that did.

6. NECESSITY/OBLIGATION

Present Simple	**MUST** **NEED TO** **HAVE TO**	The apparatus **must/needs to/ has to** be disconnected from the mains during repairs.
Present Simple negative	**NEED NOT** **DO NOT NEED TO** **DO NOT HAVE TO**	The apparatus **need not/does not need to/does not have to be** disconnected from the mains during repairs.
Past Simple	**NEEDED TO** **HAD TO**	We **needed to/had to** heat the valves before use.
Past Simple negative	**DID NOT NEED TO** **DID NOT HAVE TO** **NEED NOT HAVE**	We **did not need to/did not have to** heat the valves before use. We **need not have** heated the valves before use.

Notes:

- *We **did not need to/did not have** to heat the valves before use* does not indicate whether or not you actually heated the valves, whereas *we **need not have** heated the valves before use* implies that you did heat them, but that it wasn't necessary.
- **Must not** means 'not allowed', it doesn't mean 'not necessary'.

MODAL SENTENCES EXERCISE

Complete the sentences using *could, must, may, should, might, ought to, need to, can, have to.* Make sure you use the right tense and don't forget to use negative forms where necessary.

1. Perhaps the damage was caused by heat exposure.
   The damage ____
2. We felt sure that the damage was caused by heat exposure.
   The damage ____
3. No way was the damage caused by heat exposure.
   The damage ____
4. We don't expect heat exposure to cause any damage.
   Heat exposure ______________________________________

5. It's possible that the damage wasn't caused by heat exposure.
   The damage ____
6. I advise you to heat it.
   It ____
7. I don't think it was a good idea to expose it to heat.
   It ____

KEY

1. The damage may have been/might have been/could have been caused by heat exposure.
2. The damage must have been caused by heat exposure.
3. The damage cannot have been/could not have been caused by heat exposure.
4. Heat exposure should not cause any damage.
5. The damage may not have been/might not have been caused by heat exposure.
6. It should be heated.
7. It should not have been exposed to heat.

4.3 Writing Task: Build a Model

4.3.1 Building a model

You are now ready to begin building a model of this section, First, write a short description of what the writer is doing in each sentence in the space provided below. The Key is on the next page. Once you have tried to produce your own model you can use the Key to help you write this section of a research article when you eventually do it on your own.

GUIDELINES

You should spend 30–45 minutes on this task. If you can't think of a good description of the first sentence, choose an easier one, for example Sentence 3, and start with that. Remember that your model is only useful if it can be transferred to other Discussions/Conclusions, so don't include content words such as *stress* or you won't be able to use your model to generate Discussions/Conclusions in your own field.

Remember that one way to find out what the writer is doing in a sentence, rather than what s/he is saying, is to imagine that your computer has accidentally deleted it. What changes for you, as a reader, when it disappears? If you press another key on the computer and the sentence comes back, how does that affect the way you respond to the information?

As mentioned in previous sections, another way to figure out what the writer is doing in a sentence — rather than what s/he is saying — is to look at the grammar and vocabulary clues. What is the tense of the main verb? What is that tense normally used for? Is it the same tense as in the previous sentence? If not, why has the writer changed the tense? What words has the writer chosen to use?

Don't expect to produce a perfect model. You will modify your model when you look at the Key, and perhaps again when you compare it to the way Discussion/Conclusion sections work in your target articles.

Cognitive-behavioural stress management (CBSM) skills and quality of life in stress-related disorders ***Discussion***	**In this sentence, the writer:**
1 *Prior work has documented the effectiveness of psychosocial intervention in improving quality of life (QoL) and reducing stress in patients suffering from various disorders; Epstein,*[18] *for example, reports that orthopedic patients participating in a two-week multimedia intervention programme improved across several QoL indices, including interpersonal conflict and mental health.*	1______________

2 *However, these studies have either been short-term studies or have not focused on patients whose disorder was stress-related.* **3** *In this study we tested the extent to which an extended three-month stress management programme improved QoL among a group of patients being treated for stress-related skin disorders such as eczema.*	2____________ 3____________
4 *We found that in virtually all cases, participation in our three-month stress management programme was associated with substantial increases in the skills needed to improve QoL.* **5** *These findings extend those of Kaliom, confirming that a longer, more intensive period of stress-management training tends to produce more effective skills than when those skills are input over a shorter period via information transfer media such as leaflets and presentations (Kaliom et al., 2003).* **6** *In addition, the improvements noted in our study were unrelated to age, gender or ethnic background.* **7** *This study therefore indicates that the benefits gained from stress-management intervention may address QoL needs across a wide range of patients.*	4____________ 5____________ 6____________ 7____________
8 *Most notably, this is the first study to our knowledge to investigate the effectiveness of extended psychosocial intervention in patients whose disorder is itself thought to be stress-related.* **9** *Our results provide compelling evidence for long-term involvement with such patients and suggest that this approach appears to be effective in counteracting stress that may exacerbate the disorder.* **10** *However, some limitations are worth noting.*	8____________ 9____________ 10___________

11 *Although our hypotheses were supported statistically, the sample was not reassessed once the programme was over.* **12** *Future work should therefore include follow-up work designed to evaluate whether the skills are retained in the long term and also whether they continue to be used to improve QoL.*	11________ 12________

4.3.2 Key

In Sentence 1 *'Prior work has documented the effectiveness of psychosocial intervention in improving quality of life (QoL) and reducing stress in patients suffering from various disorders; Epstein,[18] for example, reports that orthopedic patients participating in a two-week multimedia intervention programme improved across several QoL indices, including interpersonal conflict and mental health.'* **the writer revisits previous research.**

Why should I begin the Discussion by revisiting previous research?

The start of a subsection should provide an easy entry to that subsection, and two conventional ways of doing this were discussed in the unit on Results: **offering an overview of the section by previewing the content of that subsection with some general statements** and **referring back to something from the previous sections** to link it with the new one. In the Results section, we saw that the writer may begin by summarising or referring to the method or materials used. However, it is almost impossible to give an overview of the Discussion. This is because, unlike the Methodology or Results, the Discussion covers a range of areas. As a result, many Discussions/Conclusions begin by **referring back to something from the previous sections.** This can consist of:

- revisiting the Introduction to restate the aims of the paper, important background factual information, the original prediction/theory/assumption or the problem the study was designed to solve

- revisiting the Methodology for a reminder of the rationale for the procedures followed or a summary of the procedures themselves
- revisiting the Results for a summary of the results obtained by others or by the author

Which should I choose?

One option is to begin by revisiting the most significant aspects of your work. If the most important aspect of your paper is that it provides a strong response to the gap or problem that you set up in the Introduction, fulfils your aim and/or actually solves the problem, begin by recalling that gap, aim or problem from the Introduction. If the choice of software you used or the procedure you followed or the modifications you made to existing procedures is the most important aspect of your work, begin by revisiting the Methodology. If your results are the most significant aspect of the paper because they provide confirmation of a theory or reveal something new, begin by revisiting the Results. The first sentence should not be a random choice.

You can use similar language — even similar sentences — to those in the section you have chosen to revisit. This will provide an 'echo' for the reader, and will help them recall that section. Here, the writer has responded strongly to the claims made in the literature and so uses language which is similar to the words and phrases used to state those claims in the Introduction.

In Sentence 2 *'However, these studies have either been short-term studies or have not focused on patients whose disorder was stress-related.'* **the writer revisits the Introduction to recall specific weakness in the methodology used in previous studies.**

Since the contribution of this paper is the difference between the methodology in previous research and that used here, the writer first revisits the gap/problem in the Introduction to recall the weaknesses in previous methodology which have been addressed in the present work, and then moves on to the specific differences between the methodology in the present work and that of previous work.

It is also very common to include a repeat of important background factual information at this stage in the Discussion in order to re-establish the rationale or motivation for the research. In fact background factual information is a surprisingly common feature throughout the Discussion.

In Sentence 3 '*In this study we tested the extent to which an extended three-month stress management programme improved QoL among a group of patients being treated for stress-related skin disorders such as eczema.*' **the writer revisits the methodology used in this study.**

If I revisit the Methodology here, how much detail do I need to provide?

Using the same language as in the Methodology will help the reader to remember the principles of your method, and it is common to recall significant features of your method here. However, although you can explore details of your method here, do not add new information. If information about your method is important enough to include in your research paper, it should first be given where it belongs, in the Methodology, and just recalled here.

What tense should I use to describe my methodology?

You can use the Past Simple, the Present Simple or the Present Perfect to recall your methodology or results (*In the current case HI is used/has been used/was used to define the size and shape of the turbulent structures*). If you add a short Conclusion, the Present Perfect or Present Simple are common: *We use/have used holographic data to reconstruct the three-dimensional structure.*

In Sentence 4 '*We found that in virtually all cases, participation in our three-month stress management programme was associated with substantial increases in the skills needed to improve QoL.*' **the writer revisits and summarises the results.**

Is this the same as an overview of the results?

If you provided an overview of the results early in the Results section, the content and even the structure of this sentence can be very similar. A sentence like this which summarises the results may also be needed — again using similar language and structure — in the Abstract (see the next unit on Abstracts).

So why do I need to revisit or summarise the results here too?

If you look at the diagram at the start of this section and the reasons why it is symmetrical, you can see that one of the central functions of the Discussion is to go beyond the results, to lead the reader away from a direct and narrow focus on your results towards the conclusions and broader implications or generalisations that can be drawn from those results. Summarising the results provides an appropriate starting point for that process.

In Sentence 5 '*These findings extend those of Kaliom, confirming that a longer, more intensive period of stress-management training tends to produce more effective skills than when those skills are input over a shorter period via information transfer media such as leaflets and presentations (Kaliom et al., 2003).*' **the writer shows where and how the present work fits into the research 'map' of this field.**

This is a feature of the Discussion that has not occurred anywhere else. In the short literature review in the Introduction, you gave your reader a picture of the current state of research in your field. You now need to show your reader how and where your study fits into that picture and in what way it changes or affects the research 'map' in this area. In the Discussion, it is your responsibility to make the relationship between your study and other work explicit.

What are the possible ways in which my work could fit into the picture of existing studies?

Your work may have used a different method to produce similar results, which would affect the perception of existing methods; it may confirm the results obtained in a previous study; it may contradict and therefore discredit results obtained in a previous study; it may offer a completely different or new approach or it may, as in this case, extend the results and therefore confirm the implications of previous studies. There are many ways in which your work may fit into the current research map, and these may become clearer when you look at the vocabulary for mapping later in this unit.

How do I know which studies to map my work onto? Can I mention other studies for the first time in the Discussion?

Throughout the Methodology and Results sections you have been comparing your study to existing work, and these studies are the ones you should focus on here. Although you can mention research that you have not mentioned before, it is not common to refer to a large number of studies for the first time in the Discussion. You should determine exactly which studies are affected by your work, and keep these in front of your readers at various points in your paper so that you can refer to them again in the Discussion.

In Sentence 6 '*In addition, the improvements noted in our study were unrelated to age, gender or ethnic background.*' the writer recalls an aspect of the results that represents a positive achievement or contribution of this work.

Another very important feature of the Discussion is a clear focus on the achievement or contribution of your work. Specify the nature of your achievements, using positive language that clearly presents the benefits or advantages. Don't be shy about stating your achievements. Although you are aware of what is good about the work you have done and the results you have obtained, if you do not state it explicitly, the reader may not realise the value of your achievement.

Isn't it the same as mapping?

It's similar in intention, but different in content. Mapping shows where the achievement fits into the research picture in this field, but the achievement itself is often stated separately so that the reader can see the value of what has been done and found in this study independently of how it affects the current state of knowledge.

In Sentence 7 '*This study therefore indicates that the benefits gained from stress-management intervention may address QoL needs across a wide range of patients.*' the writer focuses on the meaning and implications of the achievements in this work.

If the implications of the results were already mentioned in the Results section, isn't this repetitive?

In the unit on Results, we saw that at a late stage implications begin to be drawn from the results. It was noted that the first comment on these implications (phrases such as *suggesting that/indicating that*) was described as a pivotal move that develops the direction of the research article away from the central 'report' section towards the Discussion/Conclusion. A common mistake in Discussions is to fail to develop in this direction. It is not sufficient to present a superficial interpretation that simply re-states the results in different language. In the Discussion it is your responsibility to suggest why results occurred as they did and offer an explanation of the mechanisms behind your findings and observations. These suggestions, explanations and implications are refined, developed and discussed here.

One important difference between research writing and report writing is that the aim of research is not simply to obtain and describe results; it is to make sense of those results in the context of existing knowledge and to say something sensible and useful about their implications, *i.e.* what the results mean in that context. How do the results relate to the original question or problem? Are your results consistent with what other investigators have reported? If your results were unexpected, try to explain why. Is there another way to interpret your results? Readers need to know what they can reliably take away from your study, and it is your job to tell them. **Saying what your results *are* is the central function of the Results section; talking about what they *mean* is the central function of the Discussion.**

What if I'm not sure myself about the implications of my results?

If you look at the way implications are stated in the Discussion, you will see that the language is exactly the same as the language used to state implications in the Results. *It seems that/suggesting that/indicating that* are common here, and there is a strong reliance on modal verbs such as *may* and *could*. This is because science research never reaches an endpoint where everything is known about a particular topic; the next piece of research will refine and develop the preceding one, and so on. As a result, most science writers are careful not to make unqualified generalisations, and as you can see from the words in bold below, this writer is no exception.

4 *We found that in* **virtually** *all cases, participation in our three-month stress management programme* **was associated with substantial** *increases in the skills needed to improve QoL.* **5** *These findings extend those of Kaliom, confirming that a longer, more intensive period of stress-management training* **tends to** *produce more effective skills than when those skills are input over a shorter period via information transfer media such as leaflets and presentations (Kaliom et al., 2003).* **6** *In addition, the improvements noted in our study were unrelated to age, gender or ethnic background.* **7** *This study therefore* **indicates that** *the benefits gained from stress-management intervention* **may** *address QoL needs across a wide range of patients.*

In Sentence 8 '*Most notably, this is the first study to our knowledge to investigate the effectiveness of extended psychosocial intervention in patients whose disorder is itself thought to be stress-related.*' **the writer notes that one of the achievements or contributions of this work is its novelty.**

This sentence demonstrates that in some cases MAPPING and ACHIEVEMENT are very similar, since one of the significant achievements of this work is precisely the fact that a study of this type has not been done before.

It is difficult to be absolutely sure that no-one has ever done a particular type of study until now, so before you make such a statement you should check as thoroughly as possible. Don't rely only on the Internet. The information you get from the Internet will only be as good as your skill in looking for it, and it is unprofessional to make a mistake in a sentence like this. As we can see in Sentence 8, even after every effort has been made, the writer nevertheless includes the phrase *to our knowledge* in case a study has been overlooked accidentally.

> **In Sentence 9** *'Our results provide compelling evidence for long-term involvement with such patients and suggest that this approach appears to be effective in counteracting stress that may exacerbate the disorder.'* **the writer refines the implications of the results, including possible applications.**

Developing the implications of your work includes looking at ways in which your results might be implemented or lead to applications in the future. In this case, the results imply that *long-term involvement* should be an aspect of future treatment.

Suppose my work doesn't have any obvious applications?

Many research studies don't have obvious applications. However, it's a good idea to check in two places before you give up on the idea that your work can be applied or implemented. First, look at the beginning of your Introduction, and the first sentences and paragraphs of related work in your field. This may help you see in what way the findings in your paper can be used, because as we saw in the Introduction, the first sentence often shows in what way this research area is important or useful. Another possible source is the Discussion section of published work in this field.

It is, however, possible that the work you're involved in doesn't have a clear application at this stage — or ever. Some fields, such as engineering, are more practical than others and research can have many functions — it may be intended to clarify a theory rather than seek an applicable method. You don't need to search for or try to create applications where there are none.

> **In Sentences 10 and 11** *'However, some limitations are worth noting. Although our hypotheses were supported statistically, the sample was not reassessed once the programme was over.'* **the writer describes the limitations which should direct future research.**

This is the third time that I mention limitations — first in the Methodology, then in the Results, and now again here. How do I decide which limitations to focus on here?

The reason for mentioning the limitations of your study in the Discussion is to point out a direction for future work. You should therefore examine your study for limitations which can be addressed in future work, rather than limitations which are inherent to your research field or problems which are unlikely to be solved in the near future. Try to approach this as an invitation to the research community to continue and make progress with the topic you have investigated.

Notice that, as on previous occasions where limitations were mentioned, positive outcomes (*our hypotheses were supported statistically*) are mentioned close to the limitation in order to lessen its negative impact — in this case the positive outcome is mentioned in the same sentence.

In Sentence 12 '*Future work should therefore include follow-up work designed to evaluate whether the skills are retained in the long term and also whether they continue to be used to improve QoL.*' **the writer suggests a specific area to be addressed in future work.**

Notice the use of *therefore* in Sentence 12 to link the limitation with future research.

Why should I try to fix the direction of future work — why not encourage people to decide for themselves?

One paper will not answer all possible questions in your research area, so when you are writing the Discussion, you should keep the broader picture in mind. Where should the research go next? The best studies open up directions for research. Inviting the research community to follow your work in a specific way has many functions. First, it provides researchers with a clearly defined project, which is more attractive than a vague suggestion and therefore more likely to be carried out. Second, it encourages a line of direct continuity from your research and studies that follow directly from your own will cite your paper, which enhances the status of your study. In addition, a study which responds to the difficulties or limitations of your work may provide you with useful data for your own current and future work.

4.3.3 The model

Here are the sentence descriptions we have collected:

In Sentence 1	**the writer revisits previous research.**
In Sentence 2	**the writer revisits the Introduction to recall specific weakness in the methodology used in previous studies.**
In Sentence 3	**the writer revisits the methodology used in this study.**
In Sentence 4	**the writer revisits and summarises the results.**
In Sentence 5	**the writer shows where and how the present work fits into the research 'map' in this field.**
In Sentence 6	**the writer recalls an aspect of the results that represents a positive achievement or contribution of this work.**
In Sentence 7	**the writer focuses on the meaning and implications of the achievements in this work.**
In Sentence 8	**the writer notes that one of the achievements or contributions of this work is its novelty.**
In Sentence 9	**the writer refines the implications of the results, including possible applications.**
In Sentences 10 and 11	**the writer describes the limitations which should direct future research.**
In Sentence 12	**the writer suggests a specific area to be addressed in future work.**

We can streamline these so that our model has FOUR basic components.

1	REVISITING PREVIOUS SECTIONS SUMMARISING/REVISITING GENERAL OR KEY RESULTS
2	MAPPING (RELATIONSHIP TO EXISITING RESEARCH)

3	ACHIEVEMENT/CONTRIBUTION REFINING THE IMPLICATIONS
4	LIMITATIONS CURRENT AND FUTURE WORK APPLICATIONS

4.3.4 Testing the model

The next step is to look at the way this model works in a real Discussion (but remember it may be called 'Summary and Conclusions' instead) and in the target articles you have selected. Here are some full-length Discussions and Conclusions from real research articles. Read them through, and mark the model components (1, 2, 3 or 4) wherever you think you see them. For example, if you think the first sentence corresponds to number 1 in the model, write 1 next to it, etc.

On combining classifiers

7 CONCLUSIONS

The problem of combining classifiers which use different representations of the patterns to be classified was studied. We have developed a common theoretical framework for classifier combination and showed that many existing schemes can be considered as special cases of compound classification where all the pattern representations are used jointly to make a decision. We have demonstrated that under different assumptions and using different approximations we can derive the commonly used classifier combination schemes such as the product rule, sum rule, min rule, max rule, median rule, and majority voting. The various classifier combination schemes were compared experimentally. A surprising outcome of the comparative study was that the combination rule developed under the most restrictive assumptions — the sum rule — outperformed other classifier combinations schemes. To explain this empirical finding, we investigated the sensitivity of various schemes to estimation errors. The sensitivity analysis has shown that the sum rule is most resilient to estimation errors and this may provide a plausible explanation for its superior performance.

Phosphorus removal by chemical precipitation in a biological aerated filter

DISCUSSION

Chemical dosing onto the top of the BAF produced excellent phosphorus removal efficiencies compared to the removal obtained by biological uptake. The performance of the plant was unaffected with respect to BOD, COD, suspended solids and TKN. In contrast with previous findings, using an aluminium based reagent (Rogalla *et al.,* 1990), the nitrification process was significantly affected. The use of spent pickle liquor dosing onto an activated sludge plant for phosphorus removal was also seen to affect nitrification, especially at high doses (Bliss *et al.,*1994) although this waste product may contain contaminants toxic to nitrifying bacteria.

The resulting reduction in nitrification which occurred during chemical dosing of weight ratio 1:1.14 (P:Fe) coincided with the greatest BOD loading (1.74 kg/m^3 per day) and the highest NH4 loading (0.4 kg/m^3 per day). To achieve complete nitrification an average BOD loading of 1.6 kg/m^3 per day has been suggested (Stensel *et al.,* 1988), which was exceeded during this time. At higher BOD loadings the nitrifying bacteria may be outcompeted by the organisms responsible for carbon oxidation (Metcalf and Eddy Inc., 1979) and higher ammonia loadings can create extra pressure for the nitrifying bacteria. These conditions may also explain the increased oxygen demand. A more extensive study of the effects of iron dosing on the nitrification process may be required, although the increased BOD loading most likely accounts for the reduction in this process.

The optimum chemical dose for phosphorus removal is dependant on the EC limit imposed, the stability of the process required and the capital/running costs available. Unfortunately, specific weight ratios could not be studied for any significant period of time due to the variable concentration of incoming phosphorus. The use of Fuzzy logic systems (Bulgin, 1994), to adjust the chemical dose with respect to the incoming total phosphorus, would have removed this problem. Overall the most stable and effective weight ratio was 1:1.50 (P:Fe). This is lower than the optimum ferric chloride dose found previously (Stensel *et al.*, 1988) of 1:2.00 (P:Fe), but a comparison of the performances of iron (II) and iron (III) salts would be useful. Although this produced the most stable effluent quality, its performance was not significantly different from that produced by a dosing ratio of 1:1.25. Providing the iron (II) solution is changed regularly, because it was at the end of each period when the removal efficiency deteriorated, a chemical dose ratio of 1:1.25 should be sufficient to meet EC limits of 1 mg/litre. If, however, a limit of 2 mg/litre has to be met, the dosing ratio can be lower; 1:1.00 would be a suitable ratio. This optimum ratio is much lower than for other precipitants and processes. For example, the use of alum for phosphorus removal in aerated lagoons required a dosing ratio of 2.80:1 (weight ratio AI:P). This dose produced a 90% reduction of phosphorus on an average influent concentration of 4.80 mg/litre (Narasiah *et al.*, 1991). The addition of ferric chloride to the aeration basin of an activated sludge plant rarely achieved 0.5 mg/litre phosphorus concentrations in the effluent with weight ratios as high as 5.4:1 (Fe:P) (Wurhmann, 1968). Finally, the addition of sodium aluminate to the aeration basin of an activated sludge plant required doses of 1.7:1 (weight ratio AI:P) to produce a final effluent concentration of 1.5 mg/litre (Barth *et al.*, 1968).

In accordance with previous findings (Stensel *et al.*, 1988), chemical dosing had no significant effect on headloss during operation of the BAF, even at the higher chemical doses. Further research investigating the effects of chemical dosing on full-scale BAFs may be beneficial. The use of ferrous salts for phosphorus removal has produced good results on a pilot-scale plant. Full-scale chemical dosing with these salts has been practised for many years in Finland and Switzerland with similarly good results (Bundegaard and Tholander, 1978).

Generalized thermodynamic perturbation theory for polyatomic fluid mixtures.
I. Formulation and results for chemical potentials

VIII. CONCLUSIONS

We have derived (Sec. III) exact results relating certain background pair correlation functions in a mixture to $\beta\Delta\mu e$. This derivation makes contact with earlier results obtained by us,[13,14] and clarifies, makes rigourous, and extends the approach of Stell and Zhou.[4–7] The results hold for mixtures of arbitrary compositions and for both FHS and non-FHS systems.

We have used thermodynamic arguments to develop a general EOS for mixtures of polyatomic molecules and their constituent atoms (Sec. IV), based solely on the ideal-associated solution approximation (IASA). When the exact result for $\beta\Delta\mu e$ from Sec. III is incorporated, this theory can be seen to be a generalization of the first-order thermodynamic perturbation theory of Wertheim,[2] originally developed for tangent fused-hard-sphere mixtures. One form of this theory is based upon and requires for its implementation only thermodynamic information for the reference mixture, and the alternative form requires structural information for the reference system in the form of the background correlation function $y^*(1, 2, \ldots, m)$. Since information of the latter kind is very difficult to obtain (apart from the diatomic case), we generally advocate use of the former form of the theory. We note that the generalized theory accounts for differences in structural isomers of polymeric species, unlike other approaches.[17]

We have demonstrated that alternative implementations of the generalized EOS for fused-hard-sphere systems produce slightly different results, depending on the way in which certain quantities are calculated (Sec. V). We showed that, for bonded-hard-sphere (BHS) systems, the thermodynamically based implementation yields results identical to those obtained by using the Boublik–Nezbeda equation of state,[8] and the alternative based upon structural information yields similar, but not exact, results. This sheds light on the reason for the accuracy of results obtained by previous implementations of TPT1 for diatomic systems.[2,9,6]

We have derived expressions for the excess chemical potentials, consistent with the generalized EOS, for the components of mixtures of homonuclear polyatomic molecules and their constituent atoms (Sec. VI). Since the TPT and its generalizations have the practical drawback of requiring information concerning the properties of a reference mixture system, approximations implementable requiring only accurate knowledge of pure systems are more feasible. We have tested the results of the Lewis–Randall rule approximation[10] against those of other approximations and against some exact and near-exact results. It produces good results overall.

We have presented new and more accurate results for the individual $\beta\mu e$ and for $\beta\Delta\mu e$ for the system of tangent diatomic FHS molecules with size ratio 0.6 (System B), using both conventional NVT Monte Carlo simulations and the reaction ensemble method.[11] For this system, the simulation results show that $\beta\Delta\mu e$ is essentially independent of composition. The BN EOS was found to predict that $\beta\Delta\mu e$ is exactly independent of composition. We conjecture that this result holds for all BHS systems. This result is in agreement with the fundamental approximation of the IASA.

Optimal local discrimination of two multipartite pure states

7. Conclusion

We have demonstrated that any two multipartite pure states can be inconclusively discriminated optimally using only local operations. We have also shown that this is possible for certain mixed states and certain regimes of conclusive discrimination. We then turned our attention to finding sets of entangled states that can be recreated locally, thus allowing any global discrimination figure of merit to be achieved locally. We find that this is true for the Schmidt correlated states, and, as a consequence, this is also for any two maximally entangled states.

It would be interesting to know if there are many other states which can be locally recreated using other techniques. If this can be shown to apply to any two pure states, then we would know that two pure states can be distinguished optimally under *any* figure of merit using only local operations.

Organic vapour phase deposition: a new method for the growth of organic thin films with large optical non-linearities

4. Conclusions

In summary, we have presented a new technique, organic vapour phase deposition, for the growth of extremely pure, strongly NLO-active films of DAST via the chemical reaction of two organic vapors in a hot-wall reactor. Analysis of the films by NMR, X-ray diffraction and second harmonic generation efficiency indicates that they are chemically pure, crystalline, and exist in the monoclinic structure which has previously been shown to exhibit very large second-order non-linear optical effects. By using different reactants, and with the appropriate combinations of bubblers and solid sources, OVPD can be applied to yield thin films of many different highly polar, NLP-active organic and organometallic salts, regardless of the high vapour pressures of the materials involved. To our knowledge, growth of such compounds has not previously been possible by established methods of thin film growth. We expect this technique to open up an entirely new range of materials and numerous novel photonic device applications.

Now do the same in your target articles. We hope you obtain good confirmation of the model and can now answer the three questions at the beginning of this section:

- How do I start this section? What type of sentence should I begin with?

- What type of information should be in this section, and in what order?
- How do I end this section?

4.4 Vocabulary

In order to complete the information you need to write this section of your paper you now need to find appropriate vocabulary for each part of the model. The vocabulary in this section is taken from over 600 research articles in different fields, all of which were written by native speakers and published in science journals. Only words/phrases which appear frequently have been included; this means that the vocabulary lists contain words and phrases which are considered normal and acceptable by both writers and editors.

In the next section we will look at vocabulary for the following areas of the model, apart from:

1. REVISITING PREVIOUS SECTIONS
2. SUMMARISING/REVISITING KEY RESULTS
3. REFINING THE IMPLICATION/S

Since most of the vocabulary you need for these can be found in previous sections, there is no need here for additional vocabulary input; you can refer back to the vocabulary sections in the units on Introductions, Materials/Methods and Results to find the appropriate language. When you are REFINING THE IMPLICATIONS, use the appropriate language from the IMPLICATIONS vocabulary in the Results section and avoid conclusions and implications which are not fully supported by your data.

4. MAPPING (RELATIONSHIP TO EXISTING RESEARCH)

This includes ways to show the reader where your contribution fits into the general research picture. Phrases like *consistent with* and *provides support for* are common here.

5. ACHIEVEMENT/CONTRIBUTION

Your achievement/contribution is often stated in the Present Perfect, especially when you refer to it in the Conclusion. Sentences which begin *We have demonstrated/described/investigated/developed/shown/studied/focused on etc.* are common here.

6. LIMITATIONS/CURRENT AND FUTURE WORK

These often occur very close to each other (sometimes even in the same sentence) because the limitations of the present work provide directions and suggestions for future work. Vocabulary to describe LIMITATIONS can be found in previous sections; vocabulary for FUTURE WORK includes phrases such as *should be replicated* and *further work is needed*.

7. APPLICATIONS

Your work may not have any direct or even indirect applications, but if it does, they are mentioned here. Relevant phrases include *have potential* and *may eventually lead to*. Including APPLICATIONS lets you show the value of your work beyond the narrow aims of your specific research questions. Both APPLICATIONS and FUTURE WORK provide an interface between your research article and the rest of the world and are therefore conventional ways of ending the research article.

4.4.1 Vocabulary task

Look through the Discussions/Conclusions in this unit and in your target articles. Underline or highlight all the words and phrases that you think could be used in the seven areas above.

A full list of useful language can be found on the following pages. This includes all the words and phrases you highlighted along with some other common ones. Read through them and check the meaning of any you don't know in the dictionary. This list will be useful for many years.

4.4.2 Vocabulary for the Discussion/Conclusion

1. REVISITING PREVIOUS SECTIONS
2. SUMMARISING/REVISITING KEY RESULTS
3. REFINING THE IMPLICATIONS

When you revisit these sections, don't change the words in the sentences unnecessarily; your aim is to create an 'echo' that will remind the reader of what you said before, so repeating the same words and phrases is advantageous.

If you begin by revisiting the Materials/Methods or the Introduction, you will probably also want to summarise or revisit important results in the Discussion/Conclusion. Your results are the key evidence in support of your conclusions, and it is helpful to keep these results clearly in your reader's view.

4. MAPPING (RELATIONSHIP TO EXISTING RESEARCH)

The selection of names and studies appearing in the Discussion/Conclusion is very significant to your reader; they need to be able to group research projects together and understand how your study relates to and is different from existing research. You should identify your 'product' in terms of the research 'market'. You may also compare the work/approach of other researchers with yours in order to validate your work — or discredit theirs.

This/Our study/method/result/ approach is:	This/Our study:
analogous to comparable to compatible with consistent with identical (to) in contradiction to in contrast to in good agreement (with) in line with significantly different (to/from) the first of its kind (very/remarkably) similar (to) unlike	broadens challenges compares well (with) confirms contradicts corresponds to corroborates differs (from) extends expands goes against lends support to mirrors modifies proves provides insight into provides support for refutes supports tends to refute verify

Note: Don't forget that a simple comparative (*e.g. stronger/more accurate/ quicker etc.*) is an effective way to highlight the difference between your work and other relevant work.

Here are some examples of how these are used:

- **To the knowledge of the authors,** the data in Figs. 4–6 is the **first of its kind.**
- The results of this simulation therefore **challenge** Laskay's assumption that percentage porosity increases with increasing Mg levels.
- The GMD method provides results that **are comparable to** existing clay hydration processes.
- **Similar** films on gold nanoparticles have also been found to be liquid-like.
- Using this multi-grid solver, load information is propagated **faster** through the mesh.
- Our results are **in general agreement with** previous morphometric and DNA incorporation studies in the rat [2.6].
- Our current findings **expand** prior work.[5]
- The system described in this paper is **far less** sensitive to vibration or mechanical path changes than previous systems.
- **Unlike** McGowan, we did not identify 9-*cis* RA in the mouse lung.

5. ACHIEVEMENT/CONTRIBUTION

As you know, science writing does not generally permit the use of the exclamation mark (!), but the vocabulary used to state your achievement or contribution can still communicate that the achievement is exciting. The vocabulary list has therefore been divided into two sections; the first is a list of !-substitutes, which can be used when the achievement is very exciting, and the second is a list of slightly 'cooler' — but still positive — language.

!-substitutions

compelling	overwhelming
crucial	perfect
dramatic	powerful
excellent	remarkable
exceptional	striking
exciting	surprising

extraordinary ideal invaluable outstanding	undeniable unique unusual unprecedented vital

Positive language

accurate advantage appropriate attractive beneficial better clear comprehensive convenient convincing correct cost-effective easy effective efficient encouraging evident exact feasible flexible important low-cost novel productive realistic relevant robust	Useful verbs: assist compare well with confirm could lead to enable enhance ensure facilitate help to improve is able to offer an understanding of outperform prove provide a framework provide insight into provide the first evidence remove the need for represent a new approach to reveal rule out solve succeed in support yield

simple stable straightforward strong successful superior undeniable useful valid valuable	

Here are some examples of how these are used:

- The presence of such high levels is a **novel** finding.
- We identify **dramatically** different profiles in adult lungs.
- Our results **provide compelling evidence** that this facilitated infection.
- These preliminary results demonstrate the **feasibility** of using hologram-based RI detectors.
- Our data **rule out** the possibility that this behaviour was a result of neurological abnormality.
- The system presented here is a **cost-effective** detection protocol.
- A **straightforward** analysis procedure was presented which **enables** the **accurate** prediction of column behaviour.
- Our study **provides the framework** for future studies to assess the performance characteristics.
- We have made the **surprising** observation that Bro1-GFP focus accumulation is also pH-dependent.
- We have derived **exact** analytic expressions for the percolation threshold.
- Our results provide a **clear** distinction between the functions of the pathway proteins.

6. LIMITATIONS/CURRENT AND FUTURE RESEARCH

You will normally outline the limitations of your own work, but this is not expressed as a problem with your work, rather it provides suggestions for future work. This invitation to the research community improves the status of your work by communicating that there is much research to be done in this area.

Note that using *will* or the Present Continuous (*e.g. we will integrate/we are integrating this technique with the FEM implementations*) communicates your own intentions or work in progress; *should* is used to invite research from others (*This technique should be integrated with the FEM implementations*).

a/the need for	possible direction
at present	promising
encouraging	recommend
fruitful	remain to be (identified)
further investigations	research opportunities
further work is needed	should be explored
further work is planned	should be replicated
future work/studies should	should be validated
future work/studies will	should be verified
in future, care should be taken	starting point
in future, it is advised that…	the next stage
holds promise	urgent
interesting	worthwhile
it would be beneficial/useful	

Here are some examples of how these are used:

- Our results are **encouraging** and **should be validated** in a larger cohort of women.
- However, the neural mechanisms underlying these effects **remain to be** determined.
- This finding is **promising** and **should be explored** with other eukaryotes.
- **Future work should** focus on the efficacy of ligands synthesised in the Long group.
- An important question for **future studies** is to determine the antidepressant effects of such drugs.

7. APPLICATIONS/APPLICABILITY/IMPLEMENTATION

Research work does not always have a clear application. However, in some cases it is clear how the work can be used, particularly if your project has resulted in a device or product of some kind. In such cases, you should indicate possible applications or applicability, and in many cases this can be derived from points made earlier in the Introduction. Don't forget to use modal verbs such as *could, should* and *may.*

eventually in future soon possible	apply have potential implement lead to produce use utilise

Here are some examples of how these are used:

- Our technique **can be applied to** a wide range of simulation applications.
- The PARSEX reactor therefore could be **used** for the realistic testing of a wide range of control algorithms.
- It **should be possible**, therefore, to integrate the HOE onto a microchip.
- This approach **has potential** in areas such as fluid density measurement.
- The solution method **could be applied** without difficulty to irregularly-shaped slabs.
- Our results mean that in dipping reservoirs, compositional gradients can now **be produced** very quickly.
- This could **eventually lead to** the identification of novel biomarkers.

4.5 Writing a Discussion/Conclusion

In the next task, you will bring together and use all the information in this unit. You will write a Discussion/Conclusion according to the model, using the grammar and vocabulary you have learned, so make sure that you have both the model (Section 4.3.3) and the vocabulary (Section 4.4) in front of you.

In this unit you have seen the conventional model of the Discussion/ Conclusion and the vocabulary conventionally used has been collected. Remember that when you write, your sentence patterns should also be conventional, so use the sentence patterns you have seen in the Discussions/ Conclusions in this unit and in your target articles as models for the sentence patterns in your writing.

Follow the model exactly this time, and in future, use it to check the Discussion/Conclusion of your work so that you can be sure that the information is in an appropriate order and that you have done what your readers expect you to do in this section.

Although a model answer is provided in the Key, you should try to have your own answer checked by a native speaker of English if possible, to make sure that you are using the vocabulary correctly.

4.5.1 Write a Discussion/Conclusion

Imagine that you and your team have designed a machine which can remove chewing gum from floors and pavements by treating the gum chemically to transform it into powder and then using vacuum suction to remove it.

In the Introduction, you began by saying that chewing-gum removal is a significant environmental problem. You then provided factual information about the composition of chewing gum[1,2] and the way in which it sticks to the floor.[6] After that, you looked at existing chewing-gum removal machines[3,4] and noted that research has shown they are unable to use suction to remove gum without damaging the floor surface.[10] You referred to Gumbo *et al.*, who claimed that it was possible to use chemicals to dissolve chewing gum.[5] At the end of the Introduction you announced that you and your research team had designed a chewing gum removal machine (CGRM), which you call GumGone. GumGone sprays a non-toxic chemical onto the gum which transforms it to white powder. The machine can then remove the gum using suction without damaging the floor surface.

In the Methodology you described the design and construction of the machine. You compared your CGRM, GumGone, to two existing machines, Gumsucker[3] and Vacu-Gum.[4] You then gave details of a set of trials which you conducted to test the efficiency of the new CGRM and a further set of trials which showed the effect of gum removal on the floor surface.

In the Results section, you showed results of these trials. You compared the performance of GumGone with Gumsucker and Vacu-Gum. Your results were very good, and they can be seen in the tables below. Now write the Discussion/Conclusion.

Table 1: Gum removal as a percentage of total sample

	Gumsucker	Vacu-gum	GumGone
Wooden floor	77	73	80
Stone floor	78	78	82
Carpeted floor	56	44	79

Table 2: Floor damage/staining

	Gumsucker	Vacu-gum	GumGone
Wooden floor	minimal	minimal	none
Stone floor	significant	some	none
Carpeted floor	significant	significant	minimal

Discussion

Gum removal technology has traditionally faced the problem of achieving effective gum removal with minimal damage to floor surfaces. Existing CGRMs such as Gumsucker and Vacu-Gum use steam heat and steam injection respectively to remove gum and although both are fairly effective, the resulting staining and damage to floor surfaces, particularly carpeted floors, is often significant.[10]

In this study the design and manufacture of a novel CGRM, GumGone, is presented. GumGone reduces the gum to a dry powder using a non-toxic chemical spray and then vacuums the residue, leaving virtually no stain. In trials, GumGone removed a high percentage of gum from all floor surfaces without causing floor damage. The floor surfaces tested included carpeted floors, suggesting that this technology is likely to have considerable commercial use.

Percentage removal levels achieved using GumGone were consistently higher than for existing CGRMs on all types of floor surface. This was particularly noticeable in the case of carpeted floor, where 79% of gum was removed from a 400 m^2 area, as opposed to a maximum of 56% with existing machines. This represents a dramatic increase in the percentage amount of gum removed. Our results confirm the theory of Gumbo *et al.* that chemicals can be used to dissolve gum into dry powder and make it suitable for vacuuming.[5]

The greatest advantage over existing CGRMs, however, lies in the combination of the two technologies in a single machine. By reducing the delay period between gum treatment and gum removal, the GumGone system resulted in negligible staining of floor surfaces. This represents a new approach which removes the need for stain treatment or surface repair following gum removal.

As noted earlier, only one wattage level (400 watts of vacuum suction power) was available in the GumGone prototype. Further work is needed to determine the power level at which gum removal is maximised and floor damage remains negligible.

Unit 5 ✏ Writing the Abstract

5.1 Structure

The structure and content of the Abstract have changed in recent decades. Before on-line publication databases such as the Science Citation Index, the Abstract was printed at the top of a research article and its function was mainly to encourage the reader to continue reading the article and to facilitate that reading by providing a brief preview. The reader and the writer didn't consider the Abstract of a research article as an independent unit because it was not normally read without reference to the article itself.

The Internet has influenced the way that science research is communicated and the way that scientists access published research. Abstract databases allow scientists to search and scan the scientific literature and then decide which research articles they want to read in detail. Some readers simply want to know what is going on in their research area and may not be interested in the details; others may want to know details but are only interested in research articles which are directly relevant to their own research. However, if readers are going to actually read your research article, the Abstract now needs to persuade them to obtain a copy of it, not just encourage them to keep reading a paper they have already accessed.

Abstracts compete for attention in on-line databases. Many more people will read the title than the Abstract, and many more will read the Abstract than the whole paper. This means that however 'good' and well-written the Abstract is, it needs to have independent validity. It should make sense as a standalone, self-contained description of the research article, and readers should be able to understand the key points and results of the research even if they never see the whole article. The Abstract, in this sense, is a representation of the research article.

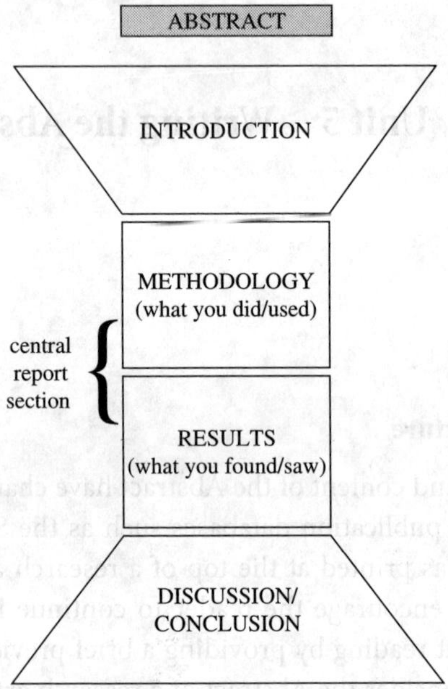

Fig. 1. The shape of a research article or thesis.

Why does the unit on Abstracts come at the end of this book rather than at the beginning?

In the first place, the style and the length of the Abstract depend on where you plan to submit it and that decision may be taken late in — or even after — the writing process. However, the most important reason for putting this unit on Abstracts at the end of the book is that you are in a better position to create an Abstract after you have finished writing the other sections of your paper. The content of the Abstract is derived from the rest of the article, not the other way around. Although you should not simply cut and paste whole sentences from the body of the article, the Abstract does not contain material which is not already in the paper. This means that you don't need to create completely new sentences; once you have decided what should go in the Abstract you can select material, including parts of sentences and phrases, from the relevant sections of the paper and adapt or modify them to meet the demands of an Abstract. This also means that the Abstract is easier to write than the rest of the paper!

Does every Abstract follow the same model?

No, and the title of the Abstract reflects this. Some are called Summary, some are called Background, some are called Abstract and others have no title at all. Most Abstracts are results-focused and there are basic similarities in all Abstracts, but there are two quite distinct models. The first model is similar to a summary, and is very structured. It deals with all the main subsections of the research article and can even have subtitles such as Background/Method/Results/Conclusions. The second model is more common, and focuses primarily on one or two aspect of the study, usually — but not always — the method and the results. Both models will be discussed here. Note that the models for an Abstract described here are appropriate for articles, papers, theses *etc.* Abstracts for conferences may not follow either of these models.

How do I know which model to choose?

This decision is based on the type of research you have done and the Guide for Authors of the journal where you want to publish your research. The decision is normally determined by the journal rather than the author. If the choice is yours, then generally speaking, the more narrow and specified your research topic, the less likely you are to use the summary format. This is because in a narrow research field, most readers already know the background. The word limit set by each journal also has a significant effect on the structure and therefore also on the content of the Abstract.

So as you can see, when we come to ask our three questions:

- How do I start the Abstract? What type of sentence should I begin with?
- What type of information should be in the Abstract, and in what order?
- How do I end this section?

You already know a lot about what the Abstract should include and in what order.

Here are examples of both models. Remember that Model 2 Abstracts are more common than Model 1.

Start by reading the Abstract below, which is an example of a structured Abstract using the summary format (Model 1). The title of the paper is: **Physical properties of petroleum reservoir fluids derived from acoustic measurements**. Don't worry if you have difficulty understanding terms such as *bubble point*. Just try to get a general understanding at this stage and familiarise yourself with the structure.

MODEL 1

Abstract: *The speed of sound in a fluid is determined by, and therefore an indicator of, the thermodynamic properties of that fluid. The aim of this study was to investigate the use of an ultrasonic cell to determine crude oil properties, in particular oil density. An ultrasonic cell was constructed to measure the speed of sound and tested in a crude oil sample. The speed of sound was measured at temperatures between 260 and 411 K at pressures up to 75 MPs. The measurements were shown to lead to an accurate determination of the bubble point of the oil. This indicates that there is a possibility of obtaining fluid density from sound speed measurements and suggests that it is possible to measure sound absorption with an ultrasonic cell to determine oil viscosity.*

Now look at an example of the second, more common, type of Abstract. The title of this paper is: **Effect of polymer coatings on drug release**.

MODEL 2

Abstract: *This study investigated the use of a novel water-soluble polymer blend as a coating to control drug release. It was found that using a blend of methylcellulose and a water-soluble copolymer significantly slowed the release rate of ibuprofen compounds in vitro and allowed for a more consistent release rate of 10–20% per hour.*

5.2 Grammar and Writing Skills

Because the Abstract is derived from the rest of the article, most of the grammar and writing skills have already been covered in previous units. The use of VERB TENSE, however, is very important in the Abstract. This section also deals with the LENGTH and LANGUAGE of the Abstract.

5.2.1 Verb tense

Verb tense is especially important in the Abstract because the strict word limit means that you may need to omit phrases that tell the reader whose work you are referring to, or what you think about your results. In this case, these can be achieved by careful and accurate use of verb tense.

Remember that the tense you use in a sentence may be grammatically correct — and therefore no editor or proofreader will notice it or draw your attention to it — but if you have not chosen the appropriate tense the sentence will not mean what you wanted it to mean and it will not have the effect you hoped it would have.

The **gap/problem** is normally in the Present Simple tense:

The main problem, however, is...
We examine why these models have difficulty with...
However, this assumption is not valid when...
This is complicated by...
However, this assessment cannot be based solely on...
Although it is known theoretically that...

When you are referring to **what the paper itself does** or **what is actually in the paper itself**, use the Present Simple tense, for example:

This paper presents a new methodology for...
In this paper we apply...
This study reports an improved design for...
In this paper we extend an existing approach to...
We consider a novel system of...
The implications for learning algorithms are discussed...
New numerical results are presented here for...

When you are referring to your **methodology**, or what you did during the research period, it is common to use the Past Simple tense, for example:

Two catalysts were examined in order to…
Samples were prepared for electron microscopy using…
A crystalliser was constructed using…
The effect of pH was investigated by means of…
The data obtained were evaluated using…
A permeameter was used to investigate…

It is also possible to use the Present Simple tense to talk about your **methodology**, especially when you are referring to calculations or equations which can be found in the paper itself:

Numerical examples are analysed in detail…
The calculated wavelengths are compared to…
Several models are created using…
The accuracy is evaluated by…
A detailed comparison is made between…
The method is illustrated on blends of homopolymers…

Results can be expressed in either the Present Simple tense, for example:

We find that oxygen reduction may occur up to 20 microns from the interface…
The model consistently underpredicts…
The ratio shifts towards…
We show that this theory also applies to…
The most accurate readings are obtained from…
We find that this does not vary…
These examples illustrate that overpotential is better described in terms of…

Or, more commonly, in the Past Simple tense, for example:

The Y-type was found to produce…
The hydrocarbons showed a marked increase in…

No dilation was observed...
This was consistent with...
Organised fibers were found after 6 weeks...
These profiles were affected by...
This finding correlated with...

but be aware that the sentence may use two different tenses. Even if the first part of the sentence is in the Past Simple tense (*We found/It was found etc.*) you can decide to put the finding/result itself or the implication of the result in the Present Simple tense if you believe it is strong enough to be considered as a fact or truth:

The experiments demonstrated there are two matrices...
It was found that proteins are produced from...
The results demonstrated that the morphology is different...
This image suggested that there is a direct relationship between...

Some of the reasons behind that choice are discussed in the unit on Introductions (Section 1.2.1) and the unit on Results (Section 3.4.2). In addition to the reasons given there, it is worth noting that the Abstract tends to present the contents of the paper in fairly direct way, not only because of the word limits imposed by editors, but also to engage the attention of the reader. This influences the decision to use the Present Simple for the results or the implications, even though those implications may have been stated in the Past Simple in the article itself.

Achievements can be expressed in the Present Perfect tense, as in the Discussion/Conclusion:

We have obtained accurate quantitative LIF measurements...
This investigation has revealed that...
We have devised a strategy which allows...
We have demonstrated the feasibility of this approach by...
A novel material has been produced which...
Three-dimensional FE predictions have confirmed that...
Considerable insight has been gained concerning...

and also in the Present Simple tense:

This process can successfully be combined with…
The framework described here is both simple and universal…
The value of our approach lies in…
This provides a powerful tool for…
This novel film is mechanically robust and is able to…
The algorithm presented here ensures that…

Applications are normally stated in the Present Simple tense:

This process is suitable for the production of…
This framework can be used to evaluate…
This approach can be applied to…
This demonstrates potential for general applicability to…
These profiles may serve as a predictor for…
This framework can be used to evaluate…

5.2.2 Length

The Abstract usually has a strict word limit. Most are between 80–150 words and are written as a single paragraph. Even longer Abstracts (150–250 words) are usually written as a single paragraph. Don't submit an Abstract that is over the word limit or it may be cut by an editor in a way that does not represent your work appropriately.

For your first draft, don't worry too much about the word limit. Once you have decided which of the two Abstract models you will use, start by including whatever you think is important, and then gradually remove words, phrases and even sentences that are not essential.

5.2.3 Language

Think of the search phrases and keywords that people looking for your work might use. Make sure that those exact words or phrases appear in your Abstract, so that they will turn up at the top of a search result listing.

The Abstract is sometimes written in a slightly less technical way than the article itself in order to attract a wider audience. This may mean that some of your readers do not know a particular technical term or acronym that you want to include. To solve this problem, you can use the acronym, abbreviation or technical term in the Abstract but you should first say what it means or stands for. For example:

Granules of hydroxyapatite (HA) were implanted.

5.3 Writing Task: Build a Model

5.3.1 Building a model

You are now ready to build a model of the Abstract by writing a short description of **what the writer is doing in each sentence** in the space provided below. This should be very easy, because all the components of the Abstract have occurred in previous subsections. As before, the Key is on the next page.

GUIDELINES

This time you will need to build two models, to cover the two types of Abstracts. You should only need to spend 10–20 minutes on this task, because the sentence types are familiar to you from previous units. Don't forget that your models are only useful if they can be transferred to other Abstracts, so don't include content words or you won't be able to use the models to generate your own Abstract.

Remember that one way to find out what the writer is doing in a sentence, rather than what s/he is saying, is to imagine that your computer has accidentally deleted it. What changes for you, as a reader, when it disappears? If you press another key on the computer and the sentence comes back, how does that affect the way you respond to the information?

As mentioned in previous sections, another way to figure out what the writer is doing is to look at the grammar and vocabulary clues. What is the tense of the main verb? What is that tense normally used for? Is it the same tense as in the previous sentence? If not, why has the writer changed the tense? What words has the writer chosen to use?

This time, you may find that you produce perfect models, but you will still probably modify them — especially the second type — when you compare them to the way Abstracts are written in your target articles.

MODEL 1

Physical properties of crude oil from acoustic measurements ***Abstract***	**In this sentence, the writer:**
1 *The speed of sound in a fluid is determined by, and therefore an indicator of, the thermodynamic properties of that fluid.* **2** *The aim of this study was to investigate the use of an ultrasonic cell to determine crude oil properties, in particular oil density.* **3** *An ultrasonic cell was constructed to measure the speed of sound and tested in a crude oil sample.* **4** *The speed of sound was measured at temperatures between 260 and 411 K at pressures up to 75 MPs.* **5** *The measurements were shown to lead to an accurate determination of the bubble point of the oil.* **6** *This indicates that there is a possibility of obtaining fluid density from sound speed measurements and suggests that it is possible to measure sound absorption with an ultrasonic cell to determine oil viscosity.*	1__________ 2__________ 3__________ 4__________ 5__________ 6__________

5.3.2 Key

In Sentence 1 *'The speed of sound in a fluid is determined by, and therefore an indicator of, the thermodynamic properties of that fluid.'* **the writer provides background factual information.**

How do I know what kind of background information to provide?

The background information that is found at the start of this type of Abstract is usually derived from the first sentences of the Introduction.

In this particular Abstract, the information provides a factual background. Other types of background may also be appropriate; for example, if your field of study is wastewater treatment or air pollution, then it may be useful to mention the political background.

How much background information should I give?

In some journals, this type of Abstract has subtitles, *i.e.* Background/Method/Results/Conclusions; if so, the number of words is usually distributed fairly evenly among the different parts, but if not, the distribution is left to the writer and the proportion of the Abstract taken by each part varies considerably. If you feel that a lot of background is *necessary to understand the Abstract itself,* combine the relevant points and summarise them in as few words as possible. The focus of an Abstract is more likely to be on the methodology or the results, so limit background information to one or two sentences.

Can I use research references in the Abstract?

Research background may be necessary, although it is rare to include actual research references. However, if your article follows directly from an existing published paper or is a major advance or contradiction of a specific work or theory, you should cite the relevant paper in the Abstract.

In Sentence 2 '*The aim of this study was to investigate the use of an ultrasonic cell to determine crude oil properties, in particular oil density.*' **the writer combines the method, the general aim and the specific aim of the study in one sentence**.

Try to combine sentences in a way that shortens the total length of the Abstract. You can reduce the number of words by combining the background information and the aim, or what this paper does and what was found, so that the sentence serves more than one purpose. Sentences such as *In order to determine x we did y* combine the aim and the method in one sentence.

In Sentences 3 and 4 '*An ultrasonic cell was constructed to measure the speed of sound and tested in a crude oil sample.* **4** *The speed of sound was measured at temperatures between 260 and 411 K at pressures up to 75 MPs.*' **the writer summarises the methodology and provides details.**

How much detail should I give?

It depends on how important the details are. In this case the methodology is the main focus of the study; the aim of the study was *to investigate the use of an ultrasonic cell* (Sentence 2). If the important contribution of your work really is in the details of the methodology, you can and should provide those details in the Abstract and you can even give those details numerically. It is quite common to find sentences which give temperatures, pressures, times, quantities thicknesses and even light-absorption data. However, in many other cases the focus of the study — and therefore of the Abstract — is not on the methodology, in which case it is given in summary form and details are reserved for the Results.

In Sentence 5 '*The measurements were shown to lead to an accurate determination of the bubble point of the oil.*' **the writer indicates the achievement of the study.**

One of the central functions of the Abstract is to emphasise new and important achievements of the study. Almost all Abstracts also include positive language at this point (*an accurate determination*) to demonstrate the value of the work.

In Sentence 6 '*This indicates that there is a possibility of obtaining fluid density from sound speed measurements and suggests that it is possible to measure sound absorption with an ultrasonic cell to determine oil viscosity.*' **the writer presents the implications of the study.**

Another important function of the Abstract is to show how the implications of the study contribute to knowledge and information in this area, and this can be derived from the aim of the study or the gap or problem the study addressed (*The aim of this study was to investigate the use of an ultrasonic cell to determine crude oil properties, in particular oil density*).

Many types of implications can be mentioned; for example, there may be implications for associated problems or for previous studies in the light of your findings.

These implications seem rather soft — is language like 'possible' really appropriate here?

It's certainly true that phrases such as *It may therefore be the case that* and other phrases that you saw in Section 3.2.4 are not common here. Results, implications and achievements are often stated quite strongly, which encourages the reader to read the rest of the article favourably and accept the conclusions. It's also true that qualifications and discussions of implications, including possible restrictions and constraints, can be left to the article itself. However, what you report in the Abstract should be consistent with what you report in the paper, and if your work represents an early stage in a breakthrough or the implications of your work are still not firm, it is appropriate to communicate this by including modal verbs (*could/might/may*) or words such as *possible*.

What do I do if there were problems with my study — do I mention those in the Abstract?

If they are really important, yes, and if so, you even briefly say what they were. It is better not to say that something *will be discussed*. The Abstract should provide/summarise the exact details of your findings. Important implications, data and findings are included, NOT left out. This includes problems, if (*but only if*) they were important, and directions for future work. Both are relatively rare in the Abstract.

MODEL 2

<table>
<tr><td>Effect of polymer coatings on drug release

Abstract

1 This paper reports the use of a novel water-soluble polymer blend as a coating to control drug release. 2 It was found that using a blend of methylcellulose and a water-soluble copolymer significantly slowed the release rate of ibuprofen compounds in vitro and allowed for a more consistent release rate of 10–20% per hour.</td><td>In this sentence, the writer:

1________

2________</td></tr>
</table>

In Sentence 1 '*This paper reports the use of a novel water-soluble polymer blend as a coating to control drug release.*' **the writer combines what the paper does** (*This paper reports*), **the method or materials used** (*water-soluble polymer blend*), **the contribution** (*novel*) **and the aim of the study** (*to control drug release*).

This shows why it is not a good idea just to copy sentences from the research article itself. The word limit in the Abstract means that you may not have space to write one sentence describing the method you used and another stating the aim of your study; you need to find a way of combining such elements. Look at these combinations:

GAP/ACHIEVEMENT

In contrast to traditional approaches to water distribution planning based on cost, the model proposed here allows issues such as quality of supply to be considered.

ACHIEVEMENT/METHOD

A substantial increase in catalyst productivity was achieved by nanofiltration-coupled catalysis.

PROBLEM/METHOD

In order to select the optimum strategy in an environment with multiple objectives, a decision-aid tool for optimal life-cycle assessment was used.

> **In Sentence 2** '*It was found that using a blend of methylcellulose and a water-soluble copolymer significantly slowed the release rate of ibuprofen compounds in vitro and allowed for a more consistent release rate of 10–20% per hour.*' **the writer refers to the method in more detail and provides numerical details of the results.**

Even when an Abstract is short it must still do almost as much work as the paper, and it should still inform potential readers whether the article is suitable for their needs. If the reader cannot decide whether to read the paper without knowing whether you used simulation, analytic models, prototype construction, or analysis of field data, you should include that. If the value and relevance of your work is that you did many experiments with various parameters rather than a single case study, you should include that information. If, as in this case, the value of the work is *a more consistent release rate of 10–20% per hour* then this should be included in the Abstract.

How much detail of the results should I give?

The results are probably the most important component of this type of Abstract, and you should be specific and give details of key results. Avoid vague words such as *small* or *better*. If you provide 'naked numbers' try and include quantitative language such as ***only*** *38%* or ***as high as*** *15%* so that the numbers cannot be misinterpreted. In this case, the writer does not simply refer to *a more consistent release rate*, the actual numerical result (*a more consistent release rate of* ***10-20% per hour***) is included. For the same reason, you should not use unclear terms such as *various methods were used* when you describe your methodology.

5.3.3 The models

Here are the sentence descriptions we have collected:

MODEL 1

In Sentence 1	**the writer provides background factual information.**
In Sentence 2	**the writer combines the method, the general aim and the specific aim of the study in one sentence.**
In Sentences 3 and 4	**the writer summarises the methodology and provides details.**
In Sentence 5	**the writer indicates the achievement of the study.**
In Sentence 6	**the writer presents the implications of the study.**

MODEL 2

In Sentence 1	**the writer combines what the paper does, the method or materials used, the contribution and the aim of the study.**
In Sentence 2	**the writer refers to the method in more detail and provides numerical details of the results.**

Rather than construct two different models, the model description given in the box below will include both types of Abstracts. We can streamline the sentence types we have collected so that the model has five basic components.

The more structured type, Model 1, typically includes the first four components in the box below in approximately the order presented; in this type of Abstract, each component tends to occur separately. These structured Abstracts occasionally include the fifth component, LIMITATIONS and/or FUTURE WORK.

Model 2 selects just two or three of the components and tends to combine components in a single sentence where possible. The components generally include RESULTS and/or ACHIEVEMENTS and frequently METHODOLOGY, but this depends on the research area and the level of specificity. A wider research focus may require BACKGROUND or AIM in the Abstract. In Model 2, the order of components is very flexible indeed — the only pattern that is generally followed is that METHODOLOGY tends to come before RESULTS.

1	BACKGROUND AIM PROBLEM WHAT THE PAPER DOES
2	METHODOLOGY/MATERIALS
3	RESULTS ACHIEVEMENT/CONTRIBUTION IMPLICATIONS
4	APPLICATIONS
5	LIMITATIONS FUTURE WORK

5.3.4 Testing the models

The next step is to look at the way this model works in some real Abstracts. Here are two Abstracts from real research articles. Read them through, and mark the model components (1, 2, 3, 4 or 5) wherever you think you see them. For example, if you think the first sentence corresponds to number 1 in the model, write 1 next to it, *etc.*

Effects of H_2O on structure of acid-catalysed SiO_2 sol-gel films

Abstract

Thin silica films were deposited on silicon wafers by the sol-gel technique, using spin coating. The sols were prepared by HCl catalysis of tetraethylorthosilicate (TEOS) diluted in ethanol, using different molar ratios, *R*, of H_2O:TEOS. The films were then baked at various temperatures, and characterised using ellipsometry, profilometry, optical scattering and infrared spectroscopy. It was found that the thickness, shrinkage, porosity and pore sizes all decrease with increasing *R*. it was also found that high water levels yield films of higher homogeneity and finer texture, and less tensile stress.

Limitations of charge-transfer models for mixed-conducting oxygen electrodes

Abstract

A framework is presented for defining charge-transfer and non-charge-transfer processes in solid state electrochemical systems. We examine why charge-transfer models have difficulty modelling non-charge-transfer effects, and walk through several examples including the ALS model for oxygen reduction on a porous mixed-conducting oxygen electrode. These examples illustrate that electrode 'overpotential' is often better described in terms of macroscopic thermodynamic gradients of chemical species. In the case of a porous mixed conducting oxygen electrode, oxygen reduction is limited by chemical reaction and diffusion, and may occur up to 20 microns from the electrochemical (charge-transfer) interface.

OPTIMIZATION AND SENSITIVITY ANALYSIS FOR MULTIRESPONSE PARAMETER ESTIMATION IN SYSTEMS OF ORDINARY DIFFERENTIAL EQUATIONS

Abstract

Methodology for the simultaneous solution of ordinary differential equations (ODEs) and associated parametric sensitivity equations using the Decoupled Direct Method (DDM) is presented with respect to its applicability to multiresponse parameter estimation for systems described by nonlinear ordinary differential equations.

The DDM is extended to provide second order sensitivity coefficients and incorporated in multiresponse parameter estimation algorithms utilizing a modified Newton scheme as well as a hybrid Newton/Gauss–Newton optimization algorithm. Significant improvements in performance are observed with use of both the second order sensitivities and hybrid optimization method. In this work, our extension of the DDM to evaluate second order sensitivities and development of new hybrid estimation techniques provide ways to minimize the well-known drawbacks normally associated with second-order optimization methods and expand the possibility of realizing their benefits, particularly for multiresponse parameter estimation in systems of ODEs.

Semi-continuous nanofiltration-coupled Heck reactions as a new approach to improve productivity of homogeneous catalysts

Abstract

Substantial increase in homogeneous catalyst productivity for a well known Heck coupling was achieved by nanofiltration-coupled catalysis. The use of nanofiltration membranes enabled catalyst separation and allowed subsequent catalyst recycle and reuse. This new technique demonstrated potential for general applicability to homogeneously catalysed organic syntheses.

Ras isoforms vary in their ability to activate Raf-1 and phosphoinositide 3-kinase

Ha-, N-, and Ki-Ras are ubiquitously expressed in mammalian cells and can all interact with the same set of effector proteins. We show here, however, that *in vivo* there are marked quantitative differences in the ability of Ki- and Ha-Ras to activate Raf-1 and phosphoinositide 3-kinase. Thus, Ki-Ras both recruits Raf-1 to the plasma membrane more efficiently than Ha-Ras and is a more potent activator of membrane-recruited Raf-1 than Ha-Ras. In contrast, Ha-Ras is a more potent activator of phosphoinositide

> 3-kinase than Ki-Ras. Interestingly, the ability of Ha-Ras to recruit Raf-1 to the plasma membrane is significantly increased when the Ha-Ras hypervariable region is shortened so that the spacing of the Ha-Ras GTPase domains from the inner surface of the plasma membrane mimics that of Ki-Ras. Importantly, these data show for the first time that the activation of different Ras isoforms can have distinct biochemical consequences for the cell. The mutation of specific Ras isoforms in different human tumors can, therefore, also be rationalized.

Now do the same in your target articles. We hope you obtain good confirmation of the model and have found the answers to the questions at the beginning of this section:

- How do I start this section? What type of sentence should I begin with?
- What type of information should be in this section, and in what order?
- How do I end this section?

5.4 Vocabulary

You already have most of the information you need to write this section of your paper because you can find the words/phrases you need in the other units of this book. However, because the Abstract needs to be understood by a wider range of people than the article itself, the Abstract tends to use simpler, more conventional language where possible. We will therefore look at the most common vocabulary in each part of the model.

The vocabulary lists in this section are taken from over 600 Abstracts in different fields, all of which were written by native speakers and published in science journals. Only words/phrases which appear frequently in this set of research articles have been included; this means that the vocabulary lists contain words and phrases which are considered normal and acceptable by both writers and editors.

In the next section we will look at typical vocabulary for all the areas of the model.

5.4.1 Vocabulary task

Look through the Abstracts in this unit and in each of your target articles. Underline or highlight all the words and phrases that you think could be used in each part of the model. You should recognise them from previous sections without too much trouble.

A full list of useful language can be found on the following pages and of course in the relevant sections in previous units. This list includes all the words and phrases you have highlighted from the Abstracts in this unit, along with others which you may have seen in your target articles.

5.4.2 Vocabulary for the Abstract

1. BACKGROUND

You can find more in Unit 1, Section 1.4.2, as well as examples of how these are used.

a number of studies	it is known that
exist(s)	it is widely accepted that
frequently	occur(s)
generally	often
is a common technique	popular
is/are assumed to	produce(s)
is/are based on	recent research
is/are determined by	recent studies
is/are influenced by	recently
is/are related to	recently-developed
it has recently been shown that	

AIM

You can find more in Unit 1, Section 1.4.2 and Unit 2, Section 2.4.2, as well as examples of how these are used.

in order to	to examine
our approach	to investigate
the aim of this study	to study
to compare	with the aim of

PROBLEM

You can find more in Unit 1, Section 1.4.2, as well as examples of how these are used.

(an) alternative approach a need for although complicated desirable difficulty disadvantage drawback essential expensive however	impractical inaccurate inconvenient it should be possible to limited not able to problem require risk time-consuming unsuccessful

WHAT THE PAPER DOES

In this study/paper/investigation we *or* We address analyse argue compare consider describe discuss emphasise examine extend introduce present propose review show	This study/paper/investigation considers describes examines extends includes presents reports reviews

Note: It is also possible to use many of these verbs with *it* or, i.e. *In this paper **it** is shown/argued that...* or in the passive, i.e. *A framework **is presented**...*

2. METHODOLOGY/MATERIALS

You can find more in Unit 2, Section 2.4.2, as well as examples of how these are used.

was/were assembled	was/were modelled
was/were calculated	was/were performed
was/were constructed	was/were recorded
was/were evaluated	was/were studied
was/were formulated	was/were treated
was/were measured	was/were used

3. RESULTS

You can find more in Unit 3, Section 3.4.2, as well as examples of how these are used.

caused	was/were achieved
decreased	was/were found
had no effect	was/were identical
increased	was/were observed
it was noted/observed that...	was/were obtained
occurred	was/were present
produced	was/were unaffected (by)
resulted in	yielded
was identified	

ACHIEVEMENT/CONTRIBUTION

You can find more in Unit 4, Section 4.4.2, as well as examples of how these are used.

accurate better consistent effective enhanced exact improved new novel significant simple suitable superior	achieve allow demonstrate ensure guarantee obtain validate compare well with for the first time in good agreement

IMPLICATIONS

You can find more in Unit 3, Section 3.4.2, as well as examples of how these are used, but remember not to use the weaker forms such as *seem to suggest* or *tend to be related to* in the Abstract.

The evidence/These results… indicate(s) that mean(s) that suggest(s) that	it is thought that we conclude that we suggest that can may

4. APPLICATIONS

You can find more in Unit 4, Section 4.4.2, as well as examples of how these are used.

applicability can be applied can be used	make it possible to potential use relevant for/in

5. LIMITATIONS and FUTURE WORK

Limitations and future work are rarely mentioned in an Abstract and then only briefly. You can find more in Unit 2, Section 2.4.2, Unit 3, Section 3.4.2, and Unit 4, Section 4.4.2, as well as examples of how these are used

a preliminary attempt not significant slightly	future directions future work

5.5 Writing an Abstract

In the next task, you will bring together and use all the information in this unit. You will write an Abstract according to the model using the grammar and vocabulary you have learned, so make sure that you have both the model (Section 5.3.3) and the vocabulary (Section 5.4) in front of you.

In this unit you have seen the two models of Abstracts and the vocabulary conventionally used has been collected. Remember that when you write, your sentence patterns should also be conventional, so use the sentence patterns of the Abstracts in this unit and in your target articles as models for the sentence patterns in your writing.

Choose one of the models, follow it exactly this time, and in future, use it to check your Abstract so that you can be sure that you have done what your readers expect you to do in this section.

Although model answers are provided in the Key, you should try to have your own answer checked by a native speaker of English if possible, to make sure that you are using the vocabulary correctly.

5.5.1 Write an Abstract

Write an Abstract for the same research that was used in Unit 4, Section 4.5.1 to write the Discussion/Conclusion. It's reprinted here in full, including the model Discussion from the Key at the end of Unit 4.

Imagine that you and your team have designed a machine which can remove chewing gum from floors and pavements by treating the gum chemically to transform it into powder and then using vacuum suction to remove it.

In the Introduction, you began by saying that chewing-gum removal is a significant environmental problem. You then provided factual information about the composition of chewing gum[1,2] and the way in which it sticks to the floor.[6] After that, you looked at existing chewing-gum removal machines[3,4] and noted that research has shown that they are unable to use suction to remove gum without damaging the floor surface.[10] You referred to Gumbo *et al.*, who claimed that it was possible to use chemicals to dissolve chewing gum.[5] At the end of the Introduction you announced that you and your research team had designed a chewing gum removal machine (CGRM), which you call GumGone. GumGone sprays a non-toxic chemical onto the gum which transforms it to white powder. The machine can then remove the gum using suction without damaging the floor surface.

In the Methodology you described the design and construction of the machine. You compared your CGRM, GumGone, to two existing machines, Gumsucker[3] and Vacu-Gum.[4] You then gave details of a set of trials which you conducted to test the efficiency of the new CGRM and a further set of trials which showed the effect on the floor surface of gum removal.

In the Results section, you showed results of these trials. You compared the performance of GumGone with Gumsucker and Vacu-Gum. Your results were very good, and they can be seen in the tables below.

Table 1: Gum removal as a percentage of total sample

	Gumsucker	Vacu-gum	GumGone
Wooden floor	77	73	80
Stone floor	78	78	82
Carpeted floor	56	44	79

Table 2: Floor damage/staining

	Gumsucker	Vacu-gum	GumGone
Wooden floor	minimal	minimal	none
Stone floor	significant	some	none
Carpeted floor	significant	significant	minimal

Discussion

Gum removal technology has traditionally faced the problem of achieving effective gum removal with minimal damage to floor surfaces. Existing CGRMs such as Gumsucker and Vacu-Gum use steam heat and steam injection respectively to remove gum and although both are fairly effective, the resulting staining and damage to floor surfaces, particularly carpeted floors, is often significant.[10]

In this study the design and manufacture of a novel CGRM, GumGone, is presented. GumGone reduces the gum to a dry powder using a non-toxic chemical spray and then vacuums the residue, leaving virtually no stain. In trials, GumGone removed a high percentage of gum from all floor surfaces without causing floor damage. The floor surfaces tested included carpeted floors, suggesting that this technology is likely to have considerable commercial use.

Percentage removal levels achieved using GumGone were consistently higher than for existing CGRMs on all types of floor surface. This was particularly noticeable in the case of carpeted floor, where 79% of gum was removed from a 400 m^2 area, as opposed to a maximum of 56% with existing machines. This represents a dramatic increase in the percentage amount of gum removed. Our results confirm the theory of Gumbo *et al.* that chemicals can be used to dissolve gum into dry powder and make it suitable for vacuuming.[5]

The greatest advantage over existing CGRMs, however, lies in the combination of the two technologies in a single machine. By reducing the delay period between gum treatment and gum removal, the GumGone system resulted in negligible staining of floor surfaces. This represents a new approach which removes the need for stain treatment or surface repair following gum removal.

As noted earlier, only one wattage level (400 watts of vacuum suction power) was available in the GumGone prototype. Further work is needed to determine the power level at which gum removal is maximised and floor damage remains negligible.

5.5.2 Key

Here are the sample answers. When you read them, think about which part of the model is represented in each sentence.

MODEL 1

Abstract

The fats and resins in chewing gum contribute to elasticity, bulk and texture but also increase staining. The aim of this study was to design a gum removal machine able to remove gum chemically with no stain residue. A machine, GumGone, was designed and constructed, which injected non-ionic detergent into gum deposits using a power spray and then immediately vacuumed the resulting powder. It was found that 1 μl of detergent achieved effective, stain-free removal over a 300 m^2 area. Performance was superior to existing systems and suggests that the delay between treatment and removal is a significant factor in staining.

MODEL 2

Abstract

This paper reports the design of a gum removal machine, GumGone, which combines non-ionic detergent treatment with immediate vacuum removal to minimise stain residue. Tests were conducted over a 300 m^2 area and removal levels of between 79% to 80% were achieved. Residual staining levels were superior to existing systems.

5.6 Creating a Title

In Section 5.5 it was stated that *Many more people will read the title than the Abstract, and many more will read the Abstract than the whole paper.* This is because the title, like the Abstract, tells readers whether or not the research article will be useful for them. A good title will attract readers and, more importantly, will attract the appropriate readers. The reverse is also true: if the title is poor the research article may not reach the appropriate audience.

I don't know how to start constructing a title.

Start by looking at your research aim or the question you were trying to answer. Try and turn the question or problem into a title. For example,

What is the difference between x and y?

becomes

A comparison of x and y
and
How does x affect y?

becomes

The effect of x on y

What is a good title?

The title should predict and describe the content of the paper as accurately as possible. If your paper is a case study, the title should reflect this:

Crack propagation in a pressurised pipe

If it is a more general survey the title should indicate this:

Crack initiation in pressurised pipes

The title should include key words that make the paper retrievable easily on search engines. It does not necessarily have to be a sentence but it should nevertheless make sense. Notice that titles of research articles don't normally use title case; they are generally written in sentence case.

There are some grammar issues that are worth noting. When you use key words in constructing the title, be careful about creating complex compound nouns. The conciseness of a compound noun is very tempting for non-native writers and English has a high level of tolerance for such nouns, but make sure that the compound noun can be understood without ambiguity. Note that the noun on the right-hand side of a compound noun is the 'real' noun and any noun or nouns to the left of it have adjectival function in the sense that they modify the right-hand noun. Also note that the relationship between the nouns that make up a compound noun may include options you had not considered:

- *an oil* ***can is a can*** *which may contain oil ....or it may be empty, but its normal use is to contain oil*
- *an oil can* ***opener is an opener*** *for cans which may contain oil*
- *an oil can opener repair* ***man is a man*** *who is able to repair cans which may contain oil*
- *an oil can opener repair man training* ***programme is a programme*** *to train men to repair openers for cans which may contain oil*
- *an oil can opener repair man training programme funding* ***problem is a problem*** *with the funding for the training programme to train men to repair openers for cans which may contain oil*

Another aspect of grammar that often causes problems — and not only in the title — is the use of prepositions such as *by, with, on, in, for*. Prepositions are not simply a type of glue to hold words together; they have a profound effect on meaning, and in the title this effect is particularly significant. The preposition *with*, for example, may mean either *using* or *having*. Evidence *for* something is evidence that tends to support or confirm that it is present or that it exists. Evidence *of* something is an actual observable sign of its presence or existence.

Filtering of code phase measures ***from*** *dual-frequency gps receivers*

is different from

Filtering of code phase measures ***in*** *dual-frequency gps receivers*

and

Sensory components controlling bacterial nitrogen assimilation

is much clearer than

Sensory components in bacterial nitrogen assimilation

Since this is such a complex area and the risk of an error in the title is so significant, it is advisable to avoid preposition-heavy structures and/or to get your title checked by a native-speaker colleague before submitting the paper for publication.

Good titles are usually concise, so it is not common to begin with phrases such as *A study of...* or *An investigation into...* They are also written in very formal English, so the use of a question mark is not common.

What can I do to make sure that readers accurately estimate the value of my paper?

If the results obtained in the study represent a significant achievement, the title may simply state the results:

> *Ras isoforms vary in their ability to activate raf-1 and phosphoinositide 3-kinase*

However, in most cases, the title is not the right place to indicate either the value of the paper or its limitations. State your title neutrally; words like *reliable* are not common, nor are modal verbs such as *may/might/could*. Be careful not to set up expectations which are not fulfilled in the paper itself; for example, if your study does not refer to all substrates/systems/reactions *etc.*, the title should specify which substrates/systems/reactions it does refer to.

Sources and Credits

The author wishes to acknowledge and thank the following, and to note that because extracts were selected to give examples of good writing, graphs and tables were not included:

Adler, S.B. (2000) Limitations of charge-transfer models for mixed-conducting oxygen electrodes. *Solid State Ionics* Vol. 135, No. 1: 603.

Burrows, P.E. *et al.* (1995). Organic vapor phase deposition: A new method for the growth of organic thin films with large optical non-linearities. *Journal of Crystal Growth* Vol. 156, Issue 1–2: 91–98.

Clark, T., Stephenson, T. and Pearce, P.A. (1997). Phosphorus removal by chemical precipitation in a biological aerated filter. *Water Research* Vol. 31, No. 10: 2557–2563.

Fardad, M.A., Yeatman, E.M., Dawnay, E.J.C., Green, M. and Horowitz, F. (1995). Effects of H_2O on structure of acid-catalysed SiO_2 sol-gel films. *Journal of Non-Crystalline Solids* 183: 261–263.

Favro, L.D. *et al.* (2000). Infrared imaging of defects heated by a sonic pulse. *Review of Scientific Instruments* Vol. 71, No. 6: 2418–2421.

Graham, N., Chu, W. and Lau, C. (2003). Observations of 2,4,6-trichlorophenol degradation by ozone. *Chemosphere* Vol. 51, Issue 4: 237–243.

Guay, M. and McLean, D.D. (1995). Optimization and sensitivity analysis for multiresponse parameter estimation in systems of ordinary differential equations. *Computers and Chemical Engineering* Vol. 19, No. 12: 1271.

Ince, B.K., Ince, O., Sallis, P.J. and Anderton, G.K. (2000). Inert COD production in a membrane anaerobic reactor treating brewery wastewater. *Water Research* Vol: 34, Issue 16: 3943–3948.

Kittler, J. *et al.* (1998). On combining classifiers. *IEEE Transactions on Pattern Analysis and Machine Intelligence Archive* Vol. 20, Issue 3: 238.

Müller, D.J. and Engel, A. (1997). The height of biomolecules measured with the atomic force microscope depends on electrostatic interactions. *Biophysical Journal* 73:1633–1644.

Nair, D. *et al.* (2001). Semi-continuous nanofiltration-coupled Heck reactions as a new approach to improve productivity of homogeneous catalysts. *Tetrahedron Letters* Vol. 42, Issue 46: 8219–8222.

Pavlovic, M.N., Arnaout, S. and Hitchings, D. (1997). Finite element modelling of sewer linings. *Computers & Structures* Vol. 63, Issue 4: 837–848.

Pendry, J.B. and MacKinnon, A. (1992). Calculation of photon dispersion relations. *London Physical Review Letters* Vol. 69, Issue 19: 2772.

Smith, W.R. *et al.* (1998). Generalized thermodynamic perturbation theory for polyatomic fluid mixtures. I. Formulation and results for chemical potentials. *Journal of Chemical Physics* Vol. 109, Issue 3: 1060–1061.

Sparks, T.H., Jeffree, E.P. and Jeffree, C.E. (2000). An examination of the relationship between flowering times and temperature at the national scale using long-term phenological records from the UK. *International Journal of Biometeorology* Vol. 44, No. 2: 82–87.

Virmani, S., Sacchi, M.F., Plenio, M.B. and Markham, D. (2001). Optimal local discrimination of two multipartite pure states. *Physics Letters A* Vol. 288, Issue 2: 62–68.

Yan, J. *et al.* (1998). Ras isoforms vary in their ability to activate Raf-1 and phosphoinositide 3-kinase. *J. Biol. Chem.* Vol. 273, Issue 37: 24052.

In addition, I would like to express my gratitude to my colleagues at the English Language Support Programme at Imperial College London for all their suggestions, and to the many students at Imperial College London who have provided research articles and input for this book over the years.

I also thank my dear children Ben, Daniel, Liora, Yoel and Alex, for their enthusiasm and support.

Useful Resources and Further Reading

Day, R. and Gastel, B. (2006). *How to Write and Publish a Scientific Paper* (6th Edition). Greenwood Press, California.

Hewings, M. (2005). *Advanced Grammar in Use.* Cambridge University Press, Cambridge.

Holtom, D. and Fisher, E. (1999). *Enjoy Writing Your Science Thesis or Dissertation!* Imperial College Press, London.

Huth, E.J. (1994). *Scientific Style and Format: The CBE Manual for Authors, Editors, and Publishers.* Cambridge University Press, Cambridge.

Jordan, R. (1990). *Academic Writing Course.* Collins ELT, London.

Krause Neufeld, J. (1987). *A Handbook for Technical Communication.* Prentice Hall, Englewood Cliffs, New Jersey.

Masters, P. (2004). *English Grammar and Technical Writing.* US State Department.

Michaelson, H. (1990). *How to Write & Publish Engineering Papers and Reports.* Oryx Press, Arizona.

Oshima, A. and Hogue, A. (1999). *Writing Academic English, Third Edition.* Longman, New York.

Skelton, J.R. and Edwards, S.J.L. (2000). The function of the discussion section in academic medical writing — Education and Debate. *British Medical Journal* 320: 1269–1270

Swales, J.M. and Feak, C.B. (1994). *Academic Writing for Graduate Students.* University of Michigan Press, Michigan.

Swales, J.M. (1990). *Genre Analysis.* Cambridge University Press, Cambridge.

Weissberg, R. and Buker, S. (1990). *Writing Up Research.* Prentice Hall, Englewood Cliffs, New Jersey.

Appendix A: Abbreviations Used in Science Writing

ABBREVIATION	FULL WORD/PHRASE	MEANING
c. (or *ca.*)	*circa*	about approximately around
cf.	*confer*	compare
et al.	*et alii*	and others
vs.	*versus*	as opposed to against in contrast to
i.e.	*id est*	that is in other words
e.g.	*exempli gratia*	for example
N.B.	*nota bene*	please note note well
p.a.	*per annum*	per year yearly

Appendix B: Prefixes Used in Science Writing

Match each prefix in Column B to the correct meaning in Column A. If two or more prefixes have the same meaning, they will be listed together. For example *poly-* and *multi-* have the same meaning (they both mean *many*) so they are listed together as 29 and 30 in Column B. Notice that the prefix *dis-* appears twice, because it has two different meanings.

COLUMN A		COLUMN B
	MEANING	**PREFIX**
________	above/more	**1. circum-**
________	after	**2. pre-**
________	again	**3. fore-**
________	against	**4. ante-**
________	apart/away	**5. anti-**
________	around	**6. contra-**
________	backwards	**7. counter-**
________	bad/badly	**8. auto-**
________	before	**9. co-**
________	between	**10. dis-**
________	change	**11. de-**
________	colour	**12. hyper-**
________	different	**13. super-**
________	equal	**14. ir-**
________	first	**15. im-**
________	half	**16. in-**
________	hundred(th)	**17. un-**
________	into/inside	**18. dis-**

____________	large/million	19. **non-**
____________	many	20. **a-**
____________	new	21. **an-**
____________	not	22. **inter-**
____________	one/single	23. **intr-**
____________	same	24. **mal-**
____________	self	25. **ill-**
____________	similar	26. **mis-**
____________	thousand	27. **neo-**
____________	thousandth	28. **post-**
____________	time	29. **poly-**
____________	too	30. **multi-**
____________	two	31. **uni-**
____________	far/distant	32. **mono-**
____________	under	33. **di-**
____________	with/together	34. **bi-**
____________	wrong	35. **semi-**

36. **re-**
37. **retro-**
38. **sub-**
39. **hypo-**
40. **infra-**
41. **hetero-**
42. **homo-**
43. **milli-**
44. **kilo-**
45. **cent-**
46. **chron-**
47. **chrom-**
48. **iso-**
49. **equi-**
50. **over-**
51. **mega-**
52. **para-**
53. **prim-**
54. **proto-**
55. **tele-**
56. **meta-**

KEY

MEANING		PREFIX	EXAMPLES
above/more	**12.** **13.**	**hyper-** **super-**	hyperactive, hyperallergenic supernatural, supersonic
after	**28.**	**post-**	postgraduate, postwar
again	**36.**	**re-**	rebuild, rewrite
against	**5.** **6.** **7.**	**anti-** **contra-** **counter-**	antioxidant, antiseptic contradict, contraindication counteract, counterpoint
apart/away	**10.** **11.**	**dis-** **de-**	disarmament, disintegrate decompose, dehydrate
around	**1.**	**circum-**	circumference, circumnavigate
backwards	**37.**	**retro-**	retroactive, retrovirus
bad/badly	**24.** **25.**	**mal-** **ill-**	malformed, malfunction ill-defined, ill-judged
before	**2.** **3.** **4.**	**pre-** **fore-** **ante-**	preexisting, pretest forecast, foresee antechamber, antenatal
between	**22.**	**inter-**	interact, interface
change	**56.**	**meta-**	metamorphosis, metastasis
colour	**47.**	**chrom-**	chromaticity, chromosome
different	**41.**	**hetero-**	heterogeneous, heterosexual
equal	**48** **49.**	**iso-** **equi-**	isometric, isosceles equidistant, equilateral
first	**53.** **54.**	**prim-** **proto-**	primitive, primordial protoplasm, prototype

MEANING		PREFIX	EXAMPLES
half	**35.**	**semi-**	semi-automatic, semicircle
hundred/th	**45.**	**cent-**	centigrade, centimetre
into/inside	**23.**	**intr-**	intravenous, introduction
large/million	**51.**	**mega-**	megabyte, megaphone
many	**29.** **30.**	**poly-** **multi-**	polysaccharide, polyvalent multicoloured, multicellular
new	**27.**	**neo-**	neonatal, neo-Darwinism
not	**14.** **15.** **16.** **17.** **18.** **19.** **20.** **21.**	**ir-** **im-** **in-** **un-** **dis-** **non-** **a-** **an-**	irrelevant, irreversible imprecise, impure inaccurate, inconsistent unbend, uncouple dissatisfied, dissimilar nonexistent, non-standard asymmetrical, atypical anaerobic, anhydrous
one/single	**31** **32**	**uni-** **mono-**	unicellular, uniform monomer, monotone
same	**42.**	**homo-**	homogeneous, homosexual
self	**8.**	**auto-**	autonomous, autopilot
similar	**52.**	**para-**	paramedic, parapsychology
thousand	**44.**	**kilo-**	kilogram, kilowatt
thousandth	**43.**	**milli-**	millisecond, millimeter
time	**46.**	**chron-**	chronological, chronometer
too	**50.**	**over-**	overheat, oversimplify

MEANING		PREFIX	EXAMPLES
two	**33.**	**di-**	dichloride, dioxide
	34.	**bi-**	bicarbonate, bisect
far/distant	**55.**	**tele-**	telemetry, telescope
under	**38.**	**sub-**	subset, subtitle
	39.	**hypo-**	hypoallergenic, hypothermia
	40.	**infra-**	infrared, infrastructure
with/ together	**9.**	**co-**	coauthor, coordinate
wrong	**26.**	**mis-**	misjudge, misread

Appendix C: Latin and Greek Singular and Plural Forms

Singular	Plural
alga	algae
analysis	analyses
antenna	antennae
appendix	appendices
axis	axes
bacterium	bacteria
basis	bases
crisis	crises
criterion	criteria
curriculum	curricula
datum	data
diagnosis	diagnoses
formula	formulae
genus	genera
hypothesis	hypotheses
index	indexes/indices
locus	loci
matrix	matrixes/matrices
medium	media/mediums
nucleus	nuclei
ovum	ova
phenomenon	phenomena
psychosis	psychoses
radius	radii

Singular	Plural
serum	sera
spectrum	spectra
stimulus	stimuli
thesis	theses
vertebra	vertebrae
vortex	vortices

Appendix D: Useful Verbs

accelerate	corroborate	imply	propose
accommodate	create	improve	prove
accompany	deal with	include	provide
account for	debate	incorporate	publish
achieve	decline	increase	purchase
acquire	decrease	indicate	put forward
adapt	define	influence	quantify
add	delay	inhibit	realise
address	demonstrate	initiate	recognise
adjust	derive	insert	recommend
adopt	describe	install	record
affect	design	interpret	reduce
allow	detect	introduce	refine
alter	determine	invert	refute
analyse	develop	investigate	regulate
apply	devise	isolate	reinforce
argue	discard	limit	relate
arise	discover	link	remain
arrange	discuss	locate	remove
assemble	display	maintain	repeat
assess	disprove	manage to	report
assist	distribute	match	represent
associate	divide	maximise	resolve
assume	drop	measure	restrict
attach	effect	minimise	retain
attempt	elicit	mirror	reveal

avoid	eliminate	miscalculate	review
bring about	employ	misjudge	revise
broaden	enable	misunderstand	rise
calculate	enhance	model	sample
carry out	ensure	modify	score
categorise	establish	monitor	select
cause	estimate	neglect	separate
challenge	evaluate	note	show
change	examine	observe	simulate
choose	exist	obtain	solve
claim	expand	occur	stabilise
classify	expect	offer	state
collect	explain	operate	study
combine	explore	optimise	substitute
compare	expose	originate	succeed
compensate	extend	outline	suggest
compute	extract	outperform	summarise
concentrate	facilitate	overcome	support
conclude	fall	overlook	test
concur	filter	peak	track
conduct	find	perform	transfer
confirm	focus on	permit	treat
connect to	formulate	plot	trigger
consider	generate	point out	undertake
consolidate	give rise to	position	use
construct	guarantee	precede	utilise
contradict	help to	predict	validate
contribute	identify	prefer	vary
control	illustrate	prepare	verify
convert	immerse	present	yield
correlate	implement	prevent	
correspond		produce	

Index of Contents

Index of Vocabulary